HUNTLEYAS *and Related Orchids*

HUNTLEYAS
and Related Orchids

Patricia A. Harding

TIMBER PRESS
Portland · London

To the orchid junkies
who helped with this book
and will help with the next one.

Page 2: *Stenotyla lendyana*. Drawing by Jane Herbst

Mention of trademark, proprietary product, or vendor does not constitute
a guarantee or warranty of the product by the publisher or the author and
does not imply its approval to the exclusion of other products or vendors.

Published in 2008 by
Timber Press, Inc.

The Haseltine Building
133 S.W. Second Avenue, Suite 450
Portland, Oregon 97204-3527
www.timberpress.com

2 The Quadrant
135 Salusbury Road
London NW6 6RJ
www.timberpress.co.uk

Printed in China

Library of Congress Cataloging-in-Publication Data

Harding, Patricia A.
    Huntleyas and related orchids/Patricia A. Harding.—1st ed.
        p. cm.
    Includes bibliographical references and index.
    ISBN-13: 978-0-88192-884-6 (alk. paper)
    1. Orchids—Identification. 2. Orchids—Classification. I. Title.
    QK495.O64H258 2008
    635.9'344—dc22   2008008877

A catalog record for this book is also available from the British Library.

Contents

Preface and Acknowledgments

How do I get myself into these things? I could talk about my deep, lifelong love of this group of orchids. I could say that I have grown many members of this group of orchids. I could say that I am the best grower of these species of plants. All this would not be true or would be at least a great exaggeration. The truth is less grand.

When I finished working with Carl Withner on *The Debatable Epidendrums*, I found myself once again with time on my hands and a desire to work on a similar task. I remembered that when I started growing orchids, I quickly became a collector, first of every orchid I could buy and then of every species I could buy or get a piece of. This practice has slowed down and even halted in the last years due to space, cost, and changes in my attitude regarding my place in the orchid community. One of my greatest frustrations in buying orchids, just after the frustration of buying a plant I already had, was buying something labeled with one name only to find that it was a synonym of a species I already possessed. There was this nagging suspicion that the plant was so labeled by the vendor to trick the buyer into buying something new.

Another frustration I had over the years is the lack of "one-stop shopping" for the identification of orchid species. The Internet will help growers more than any resource available to date, but even that is lacking and full of errors. So I have made it my small mission to pick a small group of species, sort out the taxonomic problems, write about a way to tell the species apart, and get pictures of as many of these species as possible.

Thus, with nothing to do, I started hunting around for a small group of species. I found many groups were already being studied, others too small and uninteresting to be of any notice to hobbyists in general. I both thank and curse my friend Manfred Speckmaier for suggesting this group. He encouraged me with the words, "I have lots of photographs." I had learned with the last book that photographs were not easy to come by, so Manfred's offer was a great motivator. He's not entirely evil; he did go to the herbarium in Vienna and photograph type specimens for me.

At the start of this project, I did some preliminary work on the group and came up with a rough mental estimate of 60 to 70 species. Boy, was I naïve! This book now includes approximately 200 species. When I began studying the group, I obtained as much information as I could about every single species. This approach resulted in a drawer of papers over a yard (about a meter) deep. I started sorting, typing, and trying to find patterns, hoping for synonyms to help narrow things down to a manageable level. The more I got into the group, the worse things got. Many times I thought of stopping but found that, as with the last book, telling people that I was working on a book made quitting hard and made me keep going. Pride is a cruel master, something like that.

I was fortunate that about a year into the task Mark Whitten and his crew of DNA analysts published their results on this group. With their divisions, all of a sudden I could see patterns and similarities and dissimilarities. I know they didn't do the work just for me, but I think I am their most grateful reader.

In the last few years I have been extremely fortunate to be able to meet with orchid species specialists in other countries, particularly in South America. I have been able to photograph many species, always being on the lookout for the rare and unusual ones. I also have been able, with the plant owner's permission, to bend the lip down, break it off, even tear the flower apart, and get photographs of the flower parts, so that I could, upon returning to my papers, with certainty determine the identity of a flower. Once I was sure of the characteristics of a species and what it really looked like, then I could write down for you, the reader, the traits you need to know to determine the identity of your plants.

A book has an author—me, in this case—but the real truth is that one person never writes a book like this by themselves. So many people contributed to this book with moral support, pictures, plant material, scientific material, or editing help. Most important are the individuals who gave me their opinions of problems I encountered. I list here those that helped with more than just moral support, hoping not to miss anyone. It would be too complicated to list what each one's contribution was, but no contribution was too small and each part was necessary and important.

I do have to mention someone who helped me almost daily with this book, Eric Christenson. He has seen many of these orchids over the years, but most importantly has seen and photographed or obtained photographs of the type specimens from many herbaria and has a wealth of literature and knowledge of taxonomy that made all this possible. Thank you, Eric.

Other contributors include Alejandro Abrante, Gustavo Aguirre, Roberto de Angulo, Charlie Baker, Dalton Baptista, Steve Beckendorf, Moises Béhar, Pablo Bermudez, Irene Bock and the Deutschen Orchideen-Gesellschaft (German Orchid Society), Dale Borders, Xavier Caballero, Marcos Campacci, Vitorino Paiva Castro Neto, Dave Copeland, Stig Dalström, Robert Dressler, Clare Drinkell, Günther Gerlach, Carlos Hajek, Wes Higgins, Cindy Hill, Alex Hirtz, David Hunt, Eric Hunt, Rudolf Jenny and his BibliOrchidea (a source for original publications of orchid related topics), Mario Londoño, Dick McRill, David Morris, Tilman Neudecker, Andrea Niessen and Juan Carlos Uribe, Pedro Ortíz, Ingrid Ostrander, Ron Parsons, Jay Pfahl, Andy Phillips of Andy's Orchids, Alex Portillo of Ecuagenera, Franco Pupulin, Manfred Speckmaier, Marni Turkel, Cássio van den Berg, Roberto Vásquez, Carlos Uribe Vélez, Francisco (Pacho) Villegas, Norris Williams, Gary Yong Gee, and Carol Zoltowski.

Lastly, I would like to talk about some of the limitations of myself and this book. I have seen many specimens of certain species in real life and yet I personally have seen very few herbarium specimens of these species. A classical taxonomist would have compared herbarium specimens with real-life observations as much as possible. So there, I am not a classical taxonomist in the truest sense of the word. I have taken the original description and any information I could obtain and tried to make reasonable conclusions about the species covered within. I would have liked to have been heavy-handed and "lumped" many of these species together as synonyms, but unless I was truly sure they were the same, I left the species separate, with comments about what qualities make them different.

In the future it may be that someone can determine whether actual populations are the same or different species, but at my desk this is impossible. It is a matter of opinion as to where to draw the line between a variety or form of one species and two distinct species. In the literature, there are many attempts to combine species, but when I began to look at the compilations, I often found flaws, so I mention some of these compilations in the text but generally do not incorporate them. Many species from this group have been based on one specimen without considering the total population. The authorities that follow this approach feel that putting the specimen in the record allows issues such as population and varieties to be sorted out later. With this group of plants, I am not sure that science and the orchid community are well served by this method. My hope with this book is that it will give a good foundation for the group, gathered in one place, so that if new species are to be described, authors can easily consider what has been done previously.

I received many photographs from many sources and have taken many photographs of these plants and flowers over the years. I am proud to say that I have been able to get a name to almost every flower or photograph I have considered, or at least have been able to say that it probably isn't a new species, with few exceptions. I used criteria of my own creation, that if I couldn't tell it was a new species without looking at the callus and column under a microscope or without comparing the various measurements or how the flower parts held themselves, then it wasn't worth making it a new species. The creation of a new species on a more detailed level can be done by taxonomists who have the flower in hand and who will, it is hoped, consider populations of plants. I know that this book will cause new species to come out of the woodwork (jungle or greenhouses). By not combining species as synonyms, I leave open the possibility that someone will consider one of the obscure species listed here before they give a new name to something that has been previously described and thereby create more confusion.

It is difficult to identify species by photographs alone, and yet I have been forced to do that often with this book. I hope I got them right! I have tried to not use any photograph of which I could not be sure of the identity of the flower; or, if I was unsure of the identity, I have discussed this uncertainty in the text. I have used a lot of my own photographs, not because I think I am a great photographer, but because after I photographed the specimen, I often tore the flower apart and took pictures of the parts, so that when I came home I could compare traits of the flower and be certain of the identification.

Please feel free to send me any pictures or flowers of this group you would like my opinion on. It will help if you can take two views of the flower and a third view of the lip bent down so that the underside of the column and the callus are exposed. In most cases this will not break the flower.

I want to finish this preface with a story that gives credit and applauds the work of those unsung heroes we call taxonomists. My son ran on the cross-county team in school. Though he eventually became good at it, at first he didn't do so well. My husband commented on my son's poor performance to the coach, who pointed out that though my son came in last in the race among those that ran, he came in ahead of everyone in his class in school because they didn't even try out. In the course of writing this book, I have uncovered errors made by others. It is easy to be frustrated with taxonomists especially when mistakes are made or when it means you need to change how you consider something you learned in the past. We all need to remind ourselves that at least these people made the effort to write and put down their thoughts for the record, no small feat of bravery and effort. They are much ahead of those who never make the attempt.

Introduction

In the New World is a group of genera with thin leaves arranged in a fan-shaped pattern and markedly diminished or lacking pseudobulbs. The first scientific description of a species from this group was made in 1837, when John Lindley described *Huntleya meleagris* based on a drawing of a Brazilian plant by M. Descourtilz. Lindley didn't actually see a plant till a year later when one bloomed in the collection of Messrs. Rollissons. Species from this genus and related genera have trickled in consistently since this earliest description and are still coming to the attention of the scientific community, so that now this group consists of a fairly large number of species. The group will never replace *Cattleya* or *Phalaenopsis* in the orchid trade, but several prominent orchid shows have awarded a specimen from this group "best plant" of the show.

The appeal of this group of orchids is the color of the flowers, which have a remarkable range from deep blue to waxy red and dark royal purple to almost black. Along with the striking colors are the patterns and the distinctive shapes of the flowers. There are spotted and solid-colored flowers, waxy and hairy flowers, huge and small flowers, and flat flowers and flowers with lips so curled or pouched as to be almost closed.

As most members of this group produce multiple flowers from the clump of plant growth, these plants make stunning displays. The clumps stay compact, making them easy to grow in a limited space. Growing a specimen plant with multiple growths and flowers is not a rarity but rather the norm.

The *Huntleya* Clade

A clade is a group of organisms sharing features that reflect a common ancestor. The word comes from the Greek *kladdos*, meaning branch. In the grand tree of the orchid family, *Orchidaceae*, the *Huntleya* clade is a small branch or perhaps even a twig that belongs to the New World subfamily *Epidendroi-*

deae, tribe *Maxillarieae*, subtribe *Zygopetalinae*. The subtribe is further divided into genera with obvious pseudobulbs, such as *Koellensteinia* and *Zygopetalum*, and those without obvious pseudobulbs, such as *Cryptarrhena*, *Dichaea*, and *Huntleya*.

Dressler (1993b) divided the subtribe *Zygopetalinae* into three alliances and four distinct genera, which placed the species of the *Huntleya* clade in the *Chondrorhyncha* alliance (*Huntleyinae*). He listed the traits for this group as follows: small or no pseudobulbs, conduplicate leaves, and an inflorescence with one flower. Dressler also discussed the *Chondrorhyncha* variant of the *Maxillaria* seed type, which are dust seeds or oblong dust seeds, smaller than the typical *Maxillaria* type of seed, about 250 μm long, or only three to four cells long, with marginal ridges that are weakly developed and the periclinal wall showing dense reticulate thickenings.

Molecular data showing the branching orchid tree have *Cryptarrhena* as the first branch off the main limb of *Zygopetalinae* (actually at this point the limb is more of a stick), the rest of the genera in the *Huntleya* clade, along with *Dichaea*, continue together along the stick until *Dichaea* branches off, and farther along the branch the species of *Huntleya* sit. The genera included in the *Huntleya* clade form a group that is easy to remember as vegetatively they look similar. They are markedly different vegetatively from their nearest neighbor, *Dichaea*, which makes an easy argument for not including *Dichaea* in this volume.

The genera included in this text are as follows: *Aetheorhyncha*, *Benzingia*, *Chaubardia*, *Chaubardiella*, *Chondrorhyncha*, *Chondroscaphe*, *Cochleanthes*, *Daiotyla*, *Echinorhyncha*, *Euryblema*, *Hoehneella*, *Huntleya*, *Ixyophora*, *Kefersteinia*, *Pescatorea*, *Stenia*, *Stenotyla*, and *Warczewiczella*. A number of these genera are new, the result of work by molecular taxonomists released in August 2005. Robert Dressler has been working with this group for years and doing revisions, but the 2005 studies tie much of it together. Mark Whitten, Norris Williams, Robert Dressler, Günther Gerlach, and Franco Pupulin combined in the effort to gather material for study, get the molecular analysis done and the data analyzed, and make sense of the results.

Several changes were proposed by their studies. The biggest was to divide *Chondrorhyncha* into smaller, more easily defined genera. I am sure they did not set out with this intention but only to discover the relationships of the various genera and species. Once the data were analyzed, relationships became apparent, and many previously thought relationships disappeared. The molecular data support the divisions that *Chondrorhyncha* has now undergone and actually mandate that they be done. It turns out that the species once in *Chondrorhyncha* pop up here and there all over the cladogram, with other genera in between (see Figure 1).

Though *Chondrorhyncha* species appear throughout the cladogram, they

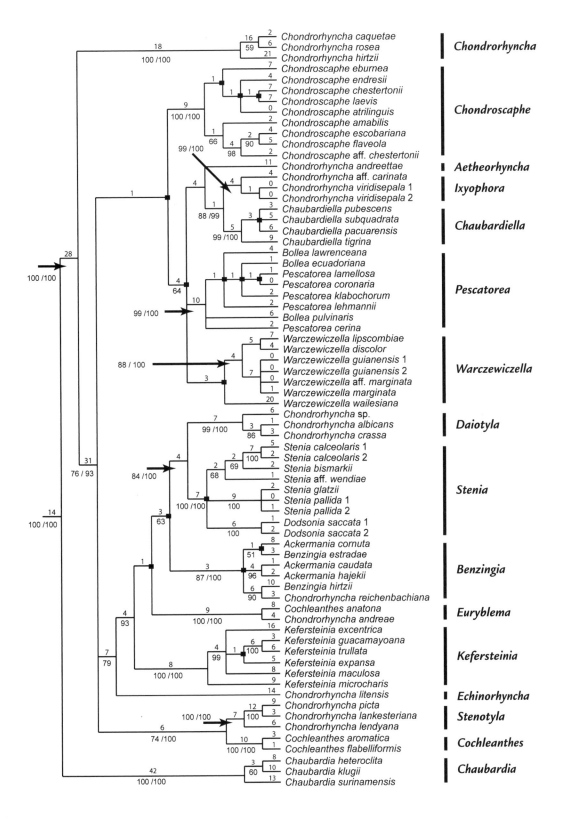

Figure 1. The *Huntleya* clade (from Whitten et al. 2005). The genus *Huntleya* is one branch above and beside this cladogram.

do so in discrete groups. The morphological traits of these groups show definite trends and similarities, so that these groups can be clearly defined. There are now new names to learn, but they provide an easier, firmer mental picture of plant attributes. The changes have not always meant more genera as some genera in the *Huntleya* clade disappeared. For example, the genus *Bollea* has now been absorbed by *Pescatorea*, and *Dodsonia* and *Ackermania* have been eliminated and included in other genera. Perhaps the biggest groan (I say this because it is so hard to spell) is that *Warczewiczella* has been resurrected, though in a different sense. Actually, it was never really gone from use, but it had fallen out of favor. Molecular data show that the genus *Cochleanthes* has only two species, and they are well separated from the other species, which by taxonomic rules of priority become *Warczewiczella*. Now it is easier to define both genera as their columns, lips, calluses, stigmas, and pollinia are more uniform in description. With this revised list of genera proposed by Whitten et al. (2005), definitions of the genera of the *Huntleya* clade are fairly clean.

Characteristics of the Group

Members of the *Huntleya* clade share many traits. The group's most noted vegetative feature is the lack of a pseudobulb. It turns out that this feature is not really lacking in all the genera and seems to be related in some genera to the size of the plant when it was collected. Nonetheless, if a pseudobulb is present, it is not prominent and one has to dig among the foliage to see it, so at first assessment it appears to be absent.

The clue to why these genera have lost the prominent pseudobulb of their ancestors and relatives originates in the habitats in which they are found in nature. These plants come from areas of constant humidity, heavy rainfall, and sufficient cloud and fog mist to ensure that they don't dry out, so they don't need the water storage system a pseudobulb would provide. Instead, their fleshy, succulent roots act in a limited way as a water storage system for the plants. Because the plants are not heavy, the roots do not need to provide a large anchor, and because moisture is plentiful, the plants don't need an extensive root system for collecting water. Most species grow on tree trunks or lower branches where the wind is not strong, and sometimes they grow in humus which is itself anchored—two more adaptations so as not to require the plant to have an extensive anchoring system.

Plants of the *Huntleya* clade have a rhizome of varying length depending on the genus. *Huntleya* species, for example, have the most prominent rhizome, often being as round as a thumb, with a good distance between vegetative plant growths. Most of the other genera have a markedly less prominent intergrowth rhizome varying to almost no rhizome between the plant growths. The leaves, inflorescence, and roots of the plant originate on the rhizome, which goes on to become the stem or pseudobulb in some genera. Roots are produced between the leaf axils, as are the inflorescences. Some plantlets of some genera have a few basilar foliaceous sheaths before the development of the actual leaf petioles and leaves.

All plants in the *Huntleya* clade are somewhat two dimensional in their growth, or more properly called alternate, or distichous, in their leaf arrangement. That is, each leaf is basally folded together lengthwise and envelopes the next younger leaf. The second leaf forms on the side of the stem opposite the first leaf, and the next leaf is again on the opposite side of the second leaf. This causes the leaf petioles to appear braided. The leaf stalk, or petiole, is always conduplicate. The actual leaf is also conduplicate with a central vein but does open in most genera to lie relatively flat. Some genera have only a strap-shaped leaf, long and narrow, but others have an elliptic leaf, widening from the basal portion to a maximum width at mid length and then narrowing again. The shape of the leaf and its thickness can be fairly characteristic and diagnostic for some of the different genera.

The inflorescence originates from the axis of the leaf sheath, or petiole, and is one-flowered. Bracts are arranged along the inflorescence and often subtend the flower and ovary. In my small orchid collection I have seen an aberrant second bud on an inflorescence, but that bud seldom develops to the point of opening and usually falls off before it fully develops. The plants tend to bloom when each growth matures. In the chapters that follow, the blooming times for the individual species in their native habitats or when they were first described are noted if they have been recorded with the type description or in other literature, but the reality is that each species blooms when it wants in cultivation, and blooming time is not diagnostic.

The flowers of the *Huntleya* clade vary in color, texture, and shape. There are the red-purples of *Pescatorea*, the brilliant red of some *Huntleya* species, the blue lips of *Chaubardia* and *Cochleanthes*, the yellows of *Chondrorhyncha* and *Kefersteinia*, and the plain old green or white flowers mixed in with several genera. I will caution the reader now and several times throughout the text that flower color often varies within a species, so use any color description as a guide, not a rule. There are alba (white) and albanistic (yellow) forms, and grades and shades of normally darkly pigmented species. Especially with *Pescatorea*, the colors are mixed up even more so because of hybrids, both manmade and natural. Textures of the flower parts vary from thick and waxy in *Huntleya* and *Pescatorea* to thin and translucent in *Kefersteinia*. Nonetheless, the flowers of this group are sturdy and fairly long lasting. The shapes of the flowers are notable for being varied. They range from the fully spread, flat, open flowers of *Huntleya* to the bell-like tubular flowers of *Ixyophora*.

Despite differences of the genera, the flowers of the clade share some characteristics. The sepals, petals, and lip are discrete; there is no evidence of fusion of these flower parts. Simple orchid flowers have the sepals, petals, and lip attached to the circular column around its base as a series of two whorls. The column in subtribe *Zygopetalinae* has a foot, or extension, at the base. The lateral sides of the column foot provide the points of insertion for the lateral sepals, and the apex of the column foot provides the point of insertion for the lip. The purpose of this foot-sepals arrangement is thought to be the formation of a mentum or nectary. If you look down the tubular lip to the back in many genera of the *Huntleya* clade, you can see an area lateral to the column foot that looks like a potential nectar tube to a pollinator. The shape of the base of the column makes the insertion of the petals occur at an oblique angle, so that the lower side of each petal is slightly longer than the upper side, as if the base of the column was trying to drag and pull the petals down onto the column foot. This structure helps form the tube created by the lip, column, and petals. Some species have small auricles, or earlike appendages, at the angles of the basal petals just at the point of insertion.

Pollination is carried out by euglossine bees, and perhaps other vectors also (van der Pijl and Dodson 1966). Long-lived male bees of the genus *Euglossa* gather perfume as a food source. The orchid species they pollinate require the bee to seek the food source (which may or may not be actually present) and retreat from the flower at a certain angle or position, so that the pollinaria is placed on a certain spot on the bee's body. Ackerman (1983) stated that *Cochleanthes* and *Huntleya* use deceit to attract nectar-seeking male and female bees of the genus *Eulaema* as the flowers lack perfume. It is not unreasonable to state that probably all genera in the *Huntleya* clade use deceit to attract pollinators as there is no nectar or food that can be gathered in these species.

Most genera in the *Huntleya* clade are thought to place the pollinia on the bee's head; however, in *Chaubardiella* the pollinaria are placed on the bee's trochanter (upper leg), and in *Kefersteinia* on the base of the bee's antennae. The mechanism for directing the insects is varied in the *Huntleya* clade, from the fimbriate callus of genus *Huntleya* and the keels on the lip of *Pescatorea* to the raised callus that is adpressed to the base of the keeled column of *Kefersteinia*. One could speculate all sorts of purposes for the lip structure of *Benzingia* and *Stenia*.

The flowers of many genera have developed tubular sepals or sepals that inroll on themselves longitudinally. In several species, if you look down the tube of the lip, you can see the tube of the rolled sepals on either side of the column foot. Though there are no data on other pollinators for these species, one could speculate that not a bee but rather something with a longer tongue could be the pollinator of at least some of these species.

Key to the Genera Based on Description of the Plants

1a. Plant large, over 30 cm high, with leaves in their natural state not fully extended . . . . go to 2

1b. Plant not large, less than 30 cm high . . . . go to 3

2a. Plant with obvious thick (1–2 cm) rhizome between growths . . . . . .*Huntleya*

2b. Plant without obvious rhizome between growths, growing in clumps . . . .*Pescatorea*

3a. Leaves straplike, not much wider than the basal petiole or leaf sheath. . . . . go to 4

3b. Leaves obviously widening mid length by at least double the leaf sheath width. . . . . go to 7

4a. Leaves thin and papery except for midvein . . . . *Chondroscaphe* or *Kefersteinia*

4b. Leaves thick, not papery . . . . go to 5

5a. Leaf sheath with red spots . . . . *Euryblema*

5b. Leaf sheath green without red spots. . . . . go to 6

6a. Leaves green, not papillose on surface . . . .*Aetheorhyncha, Chaubardia, Cochleanthes, Hoehneella,* or *Warczewiczella*

6b. Leaves glaucous gray-green, papillose on surface. . . . . *Benzingia*

7a. Leaves elliptic. . . . . . .*Chondrorhyncha, Echinorhyncha, Stenotyla,* or *Kefersteinia*

7b. Leaves obovate. . . . . go to 8

8a. Leaves green, not papillose on the surface. . . . .*Chaubardiella, Daiotyla, Ixyophora,* or *Stenia*

8b. Leaves glaucous gray-green, papillose on surface. . . . . *Benzingia*

Key to the Genera Based on Description of the Flowers

1a. Lip generally concave but not forming a tube or cup, lip lateral edges generally not clasping or curling up toward the column (except in a few species of *Pescatorea*) . . . . go to 2

1b. Lip lateral edges clasping or curling up toward the column and forming a tube. . . . . go to 7

2a. Callus with fimbriae or teeth . . . . go to 3

2b. Callus without fimbriae . . . . go to 5

3a. Column apex crested . . . . *Huntleya*

3b. Column apex smooth. . . . . go to 4

4a. Callus teeth long and fimbriate. . . . . *Chaubardia*

4b. Callus teeth merely raised, not long and fimbriate . . . . *Hoehneella*

5a. Callus concave, sometimes raised after the concavity, callus not ending in a sulcus. . . . . *Chaubardiella*

5b. Callus concave, then markedly convex, callus not extending to midlobe but ending in a sulcus . . . . go to 6

6a. Callus concave, then markedly convex for at least three times the length of the concavity, keels not becoming veins of lip . . . . *Pescatorea*

6b. Callus concave, then convex for about equal lengths, keels becoming veins of lip . . . . *Cochleanthes*

7a. Dorsal sepal and petals spreading, at right angles to column, not covering the column . . . . go to 8

7b. Dorsal sepal, petals, and lip lying parallel with the column at least one-third of length . . . . go to 12

8a. Lip deeply pouched with apical end pinched and nearly closed. . . . . . *Stenia*

8b. Lip not deeply pouched with apical end not pinched and nearly closed. . . .
. . . . go to 9

9a. Callus two lamellae or lobes . . . . *Kefersteinia*

9b. Callus a plate or series of keels . . . . go to 10

10a. Callus a simple transverse plate not composed entirely of keels. . . *Benzingia*

10b. Callus composed entirely of keels across its width. . . . . go to 11

11a. Callus composed of keels that radiate out from base to form a raised . . . . plate . . . . *Warczewiczella*

11b. Callus a series of plates that become the veins of the apical lip . . . .
. . . . *Cochleanthes*

12a. Callus two lobes or two mounds . . . . go to 13

12b. Callus a plate or a series of keels that form a transverse plate . . . . . go to 15

13a. Callus two lamellae with central sulcus with a second thickening of lip apical to the callus . . . . *Chondroscaphe*

13b. Callus not oblong, no second thickening of the lip . . . . go to 14

14a. Callus medial, having two lobules atop a stalk or a small plate of two lamellae, callus not extending to lateral margins of lip . . . . *Kefersteinia*

14b. Callus two easily visible broad large thick mounds that extend laterally. . .
. . . . *Daiotyla*

15a. Lip pubescent . . . . go to 16

15b. Lip not pubescent . . . . go to 17

16a. Lip pubescent on the apical flare . . . . *Aetheorhyncha*

16b. Lip not pubescent on the apical portion. . . . . *Euryblema*

17a. Column bearing two or more bristly sea urchinlike appendages on the underside, clinandrium on ventral side of column so that anther is not seen from the front . . . . *Echinorhyncha*

17b. Column not bearing sea urchinlike appendages on the underside, clinandrium at least partially apical . . . . go to 18

18a. Callus without keels, plate medial and basal only . . . . *Stenotyla*

18b. Callus with keel or keels. . . . . go to 19

19a. Central keel not most prominent. . . . . *Ixyophora*

19b. Central keel most prominent (highest or thickest). . . . . go to 20

20a. Callus medial, not extending to lateral edge of lip. . . . . *Chondrorhyncha*

20b. Callus extending to lateral edge of lip. . . . . *Benzingia*

Cultivation

Many species of the *Huntleya* clade are cultivated, though some genera are easier and therefore more common in cultivation than others. These plants can be so rewarding to grow and yet frustrating at the same time. Many growers have stories of large plants that were growing well but suddenly went downhill, with all attempts to rescue the plant failing. This story is particularly told in instances after a grower has divided a large plant and lost every division. Still, people who grow these orchids tend to really enjoy their plants and are eager to try again with different clones or species.

The plants seem to be habit forming among collectors who are successful in growing them. Part of the reason for this is, excluding the large *Huntleya* species, the plants are rather small to moderate in size and stay in their container, forming a nice clump. They bloom off and on over several months with a rest in the shorter day-length months or in cooler weather. Once you are successful with one, you can carry that success over into other species and genera in the group.

So how do you grow them? I think you first need to remember where the plants grow in nature, generally in the low levels of the forest, growing on the lowest 10 feet (3 meters) of a tree trunk or occasionally terrestrially in the leaf duff. The plants are often in rather dense forest, and their host tree does not reach the upper levels of the forest but lives in the shade of other trees. Only occasionally do most of the species receive direct light, which is consistent with their thin leaves, a characteristic often seen in plants that grow in low light.

Plants of this clade are found in areas of high humidity, almost daily rain, and with night temperature cooling so that humidity condenses on the plant and its roots. The plants often grow at right angles to the tree trunk, with the leaves (and the crown of the plant) hanging down so that water runs off the plant rather than remaining in its crevices. The roots can be almost 3 feet (1 meter) long, spreading both up and down the tree trunk.

With these general growing conditions in mind, prospective growers also have to consider variations in elevation (and therefore temperatures) and amount of rainfall in the different areas and regions. Many species of this group were described as coming from places that are extremely uncomfortable for humans, mostly because of humidity, heat, and mud. To cultivate these plants, you need to reproduce the conditions of their origin (where the host tree is) to the best of your ability.

When you obtain a plant of this group, get recommendations or details on how the previous grower was growing it and try to reproduce those conditions. Knowing where a plant comes from and its habitat may help, but since many of these species tolerate a variety of conditions, going with what works for another grower can't hurt.

Plants in the *Huntleya* clade tend to behave differently in cultivation than they do in nature. In nature the roots are long and spread out, whereas in successful cultivation they form a root ball, or a mop of roots coming from the base of the plant that fills the container in which the plant is growing. In cultivation the plants tend to grow upright rather than pendent. This makes lush foliage plants with a circle of flowers around the base of the plant, usually having the flowers just below the outermost leaf tips of the plant.

Plants of this clade are not able to adapt to harsh conditions because they lack pseudobulbs with reserve water and energy. How much water and humidity a particular species needs depends on the species. In general, *Huntleya* and *Pescatorea* species are found in humid and rainy areas while the other genera tend to come from drier climates and hence do not need as much water available to their roots at all times. Because all members of the clade come from regions where dew forms most nights, you must consider that they never want to dry out completely.

Huntleya species are perhaps the hardest to grow, probably because they want warm wet environments, which also promote bacteria and fungus. I know many people around the world who have tried growing them and the plants have died. I have grown a *Huntleya meleagris* for nearly twenty years. I got the plant from a grower's estate sale, but I do not know from which county or climate the plant came. I grow it in my intermediate to warm room, with a low temperature of 56°F (13°C), a high of 110°F (43°C). The plant receives rain water and a general fertilizer through an overhead watering system. I keep the base of the pot in a shallow tray of water. This plant grows well and blooms regularly for me. I have divided it, and the divisions do well for others, who will admit it is the only *Huntleya* they have ever been able to grow. I have seen wonderfully grown huntleyas at shows in Colombia and Peru and in the greenhouses at Ecuagenera in Ecuador, so I know they can be cultivated out of their natural habitat. The few successful growers of these species will say the plants are difficult and often take years in cultivation to initiate blooms. Successful growers of *Huntleya* are scattered here and there, not just in South America, but most grow just one or two plants and report that once the plant is established it seems to be more resistant to adversity (if it doesn't die, it lives).

Pescatorea and *Chaubardia*, in contrast, are very easy to grow, but once again you have to give them what they want. My pescatoreas grow right beside the *Huntleya* described above. I grow them in baskets or paper pots with the base of the pot or basket in a tray of water. Plastic will also work, but the plants tend to grow better for me if the roots can also breathe. I've seen individuals with lots of pescatoreas grow them both under and on top of benches, in cool and warm conditions depending on where the greenhouse is. All growers give them constant moisture with frequent watering.

Kefersteinia species grow in many climates and treating all the same could be improper, though most growers with several species say they treat all their *Kefersteinia* plants the same. The plants that don't thrive are no longer part of their collection. *Kefersteinia* species in nature are found in places that are generally high in humidity and have good air movement. When I asked for recommendations from experienced growers of this genus, I got many different responses. These growers use various kinds of potting media from sphagnum moss to bark to mounts, but all recommend that the potting media never get stale. This genus needs to never be really dry, but also to never be soggy wet for long. Potting material needs to be loose about the roots so that the roots can have air and don't have to aggressively dive through the media. Some growers use ceramic pots and others use plastic pots, depending on their ambient humidity. I have used paper pots with *Kefersteinia* and that has worked well also, but the pots tended to dry out too quickly, and setting the pot in a tray of water was too wet for the plant, so I no longer use paper pots for them.

The remaining genera in the *Huntleya* clade grow under conditions similar to those for *Kefersteinia*, but temperature is more critical for some of them. Consult the person from whom you get a plant for information about growing it. For the most part, I have seen these remaining genera grown much drier than *Kefersteinia*. Some growers even recommended using mounts with some species.

Chaubariella is a bit problematic in that its inflorescence is so pendent as to be adpressed to the substrate when grown in a pot. If you wish to grow it in a pot, make a mound about 1½ inches (3–4 cm) high of the substrate (I used coarse peat). This will enable the plant to grow out of the top of the mound and the inflorescence to lie on the edges of the mound so that the flower is well displayed and not hidden by the foliage.

Several growers reported that water quality was probably a critical factor in this group of plants. They noticed a decline in their plants with an adverse change in water quality. These growers recommended watching the parts per million of salt in the water and the pH, and avoiding water with added chlorine. Adding fertilizer requires watching for pH changes and though these plants are not heavy feeders, they will require some well-balanced fertilizer.

Experienced growers also recommended low light and fair air movement, though in practice light levels for these genera are not all that important. I have seen plants grown commercially in bright diffuse light and beneath the bench, both with good results, sometimes by the same grower and with the same species. I don't believe the plants adapt. Rather, if moisture is adequate, plants can handle more light, but they don't require more light to grow well.

Pests are not a problem for these plants in my experience. Pests prefer other plants in your greenhouse over plants from this group. Snails and slugs are a problem, but not any more for these genera than for other greenhouse plants.

Bacterial and fungal rot are problems due to the constant moisture and humidity of the growing conditions. These plants get a rot that seems to affect the new growth. Once the plant has this rot, it is difficult to control without the use of chemicals and it will eventually kill the plant if left untreated. This rot presents itself as a brown area usually near the end of the leaves that will progress down to the base of the leaf. If you cut off the brown area, it will reappear later farther down on the leaf. If you remove the leaf, the plant will not succumb, but the brown will appear on the next new leaf of that growth and eventually on the leaves of the neighboring new growth, as if the rot is systemic. The rot doesn't necessarily kill the plant if you remove the affected area, but the plant is left with only the outer sheaths, which of course just keeps it in limbo. The use of the fungicide Subdue (metalaxyl) is reported by successful growers to control this rot.

If you are successful in growing these plants, then eventually you will need to divide them. The plants do better when they are divided with three or four growths in a clump rather than one or two growths. They are sold often as two-growth divisions, especially as imports, and frequently the stress of the trip dries them out and they are lost. These plants have minimal storage systems, which are compromised simply by dividing the plants. It is best to take a multigrowth plant, say with 7 to 10 growths, and divide such a plant in half or thirds, as you will be more assured that the parts will succeed. As with other orchids it is probably best to divide huntleyas and their relatives when they are just starting new roots, but these species can be in continual growth and most are sporadic bloomers so picking a season to repot and divide would be difficult and perhaps not really that important.

Aetheorhyncha Dressler, *Lankesteriana* 5(2): 94. 2005.

Aetheorhyncha is a genus with a single species once placed in *Chondrorhyncha*. This species, though lovely, was a problem for people working on *Chondrorhyncha*, as it never really seemed to fit. In gross features—the tubular lip, the curved sepals—it resembles *Chondrorhyncha*, but the finer aspects of the floral parts were not easily made to fit within the older definitions of *Chondrorhyncha*. In 2005, supported by molecular data, Robert Dressler placed this species in its own genus.

Aetheorhyncha is weakly supported by molecular data as a sister group to *Ixyophora* but does not fit within any other group including its closely related molecular neighbors. *Aetheorhyncha* shares some morphological similarities with *Chondroscaphe* but is more remote genetically from *Chondroscaphe* than *Chaubardiella*, *Ixyophora*, *Pescatorea*, and *Warczewiczella*. When the morphology of these four genera is compared to that of *Aetheorhyncha*, it becomes clear that *Aetheorhyncha* is a separate genus.

The single species of *Aetheorhyncha* has the characteristic imbricating basal foliar sheaths of the *Huntleya* clade, a feature associated with the old *Chondrorhyncha*. The inflorescence holds the flower in a semiupright position, clear of the foliage often with the apical end uppermost. The flowers are bell-shaped and tubular. The dorsal sepal is erect with its outer margins rolling in slightly. The lateral sepals inroll along their length, forming a tube, and reflex back or laterally. The petals are lightly pleated apically and obliquely attach to the column basally. The petals and dorsal sepal form a tube with the lip, with just the apical margins of the lip and petals bending back at right angles. The lip forms a complete tube by itself, almost entirely encircling the column. The blade of the lip is pubescent or warty apical to the callus. The lip has a strong median keel extending from the base of the lip to the mid callus. The callus is a narrow flat plate with two blunt, broad apical points. (The original description says this is a two-lobed callus, but the lobes are not separate and are very flat.) The callus extends apically to two-thirds the length of the lip. The column does not extend apically over the clinandrium.

Etymology

Greek *aethes*, strange or different, and *rhynchos*, snout or muzzle, suggesting a relationship to *Chondrorhyncha* but not describing a specific flower feature.

List of the Species of *Aetheorhyncha*

*Aetheorhyncha andreettae** (type species of *Aetheorhyncha*)

* Molecular sampling confirms placement of this species within the genus *Aetheorhyncha*.

Key to the Species of *Aetheorhyncha*

Flower white with red spots in lip . . . .*Aetheorhyncha andreettae*

Aetheorhyncha andreettae (Jenny)

Dressler, *Lankesteriana* 5(2): 95. 2005. Basionym: *Chondrorhyncha andreettae* Jenny, *Orchidee (Hamburg)* 40(3): 92. 1989. Type: Ecuador, Cutucu, 900 m, collected by Angel Andreetta 1987 (holotype: G).

DESCRIPTION The sepals and petals are white and the lip is white with a dark yellow throat marked with red spots. The colors are vibrant and the hairy lip surface has a crystalline texture. The lip is broader at the base when spread, crispate on the margin, and flared for the apical third of the lip length. The callus is a bilobed plate with a central keel joining the two lobes at midline; each flat lobe is obliquely notched at its apex. The callus extends to just over half the lip length. The reflexed apical lip is pilose. Though not shown in Figure 2, the column rotates or twists along its length so that the ventral column does not lie lowermost but lies lateral to the column. PLATE 1, FIGURE 2.

COMMENT An illustration in *Icones Plantarum Tropicarum* (1991: ser. 2, plate 413), incorrectly labeled as *Chondrorhyncha andreettae*, shows *C. paucimaculata*. An illustration in *Icones Orchidacearum Peruviarum* (plate 16) displays a plant with a callus and a column consistent with the genus *Chondroscaphe* and is probably *C. plicata*.

MEASUREMENTS Leaves 12–16 cm long, 3.5–4.0 cm wide; inflorescence 5 cm long; dorsal sepal 2.0–2.1 cm long, 0.6–0.7 cm wide; lateral sepals 2.5–2.6 cm long, 0.7–0.9 cm wide; petals 2.2–2.5 cm long, 1.0–1.2 cm wide; lip 1.9–2.5 cm long, 2.0–2.4 cm wide; column 1.9–2.0 cm long.

ETYMOLOGY Named for Angel Andreetta of Ecuador, honoring his knowledge and interest in orchids.

DISTRIBUTION AND HABITAT Known from eastern Ecuador, Colombia, and Peru at 900–1850 m, in wet montane forest.

FLOWERING TIME November to April.

Figure 2. *Aetheorhyncha andreettae* and lip. Drawing by Jane Herbst

Benzingia Dodson, *Icon. Pl. Trop.*, ser. 2, 5: plate 406. 1989.

Calaway Dodson first published the genus *Benzingia* in 1989 when describing a new species, *B. hirtzii*. The International Code of Botanical Nomenclature requires that, in the establishment of a new genus, a type be designated, and some felt that this had not been done, as pointed out by Rosemary Davies in *Index Kewensis* in 1994. Dodson and Gustavo Romero (1995) subsequently corrected this oversight in a later article, explicitly designating *B. hirtzii* as the type species.

Eric Christenson (pers. comm.) points out that *Benzingia* was published with *B. hirtzii* as

> gen. et sp. nov., meaning the description covers both the new genus and the species. Thus, *Benzingia hirtzii* is by literal definition the type of the genus. The "revalidation" by Dodson and Romero (1995) is completely unnecessary. You can't have an identical genus and species description and then argue that a second species could be the type species for the genus.

In 2005, Whitten et al. combined *Ackermania*, *Benzingia*, and *Chondrorhyncha reichenbachiana* into one genus based on what is considered strong molecular data and morphological characteristics. The name *Benzingia* has priority as the name for the group, and the type remains *B. hirtzii*.

Benzingia species are pendent epiphytic plants. The rhizome is short, so the plants grow as clumps. There is no pseudobulb, but the short stem is surrounded by imbricating distichous leaves bearing membranous sheaths. These sheaths articulate with the narrow leaf base. The leaves of *Benzingia* widen minimally from the conduplicate base and are pointed. The leaves of most species are glaucous gray-green and pendent. The upper surface of the leaf cells is papillose, which gives the leaf surface a sparkling appearance. In other genera the leaf surface is smooth.

One inflorescence is produced per leaf sheath and the inflorescence is pendent, though in a plant grown upright the inflorescence may appear erect or laxly erect. The ovary is geniculate, resulting in the flower facing forward in most species. The flowers are both resupinate and nonresupinate depending on species. The sepals and petals are spreading and erect, not forming a tube, excepting *Benzingia reichenbachiana* and to a lesser degree *B. hajekii* and *B. thienii*. The lips are variable and can be minimally concave or shallowly saccate to deeply tubular. The callus is broad and flat, with a central keel that is more palpable than visible. The columns are fairly straight and are widest at the stigma or just below it. The wide portion of the column is flat without wings, but if the lateral edges of the column do fold downward they do so minimally in comparison to other genera. The clinandrium is more or less on the ventral side of the column, though this is variable in the drawings I have seen. The column foot is elongate and generally half the length of the entire column. The four pollinia hold the smaller pair innermost. The pollinia are attached to a rectangular stipe, which is attached to a cordiform viscidium.

Etymology
Named in honor of David Benzing, who participated in the collection of the type.

List of the Species of *Benzingia*
*Benzingia caudata**
*Benzingia cornuta**
*Benzingia estradae**
*Benzingia hajekii**
*Benzingia hirtzii** (type species of *Benzingia*)
Benzingia jarae
Benzingia palorae
*Benzingia reichenbachiana**
Benzingia thienii
* Molecular sampling confirms placement of this species within the genus
  Benzingia.

Key to the Species of *Benzingia*
1a. Lateral sepals not reflexed backward, lip not tubular, oblong, and open at apex . . . . go to 2
1b. Lateral sepals reflexed backward, lip tubular, oblong. . . . . go to 8
2a. Lip with one lobe or if multiple lobes those lobes are obscure, lip concave and open, callus easily seen from flower front . . . . go to 3
2b. Lip with more than one lobe and saclike. . . . . go to 6
3a. Lip edge not reflexed. . . . . go to 4
3b. Lip edge reflexed or deflexed. . . . . go to 5

4a. Central keel on apical lip becoming apical point or horn, lateral edges of lip at or above column edge . . . .*B. hajekii*

4b. No central keel on lip, simple point on lip almost a bump on apical end of lip, lateral edges of lip below column . . . .*B. cornuta*

5a. Lip round with no point at apex, flower heavily spotted maroon . . . .*B. hirtzii*

5b. Lip with apical point, flower yellow with only maroon at base of lip and sepals and petals . . . . *B. estradae*

6a. Lip margins revolute, column foot equal length with rest of column . . . . *B. palorae*

6b. Lip margins not revolute, column foot twice as long as the rest of the column . . . . go to 7

7a. Lip with smooth lateral margins . . . . *B. caudata*

7b. Lip with lateral margins forming toothlike points . . . . *B. jarae*

8a. Lip open at apex, deflexing slightly . . . .*B. reichenbachiana*

8b. Lip apical margins inroll . . . .*B. thienii*

Benzingia caudata (J. D. Ackerman) Dressler,

Lankesteriana 5(2): 93. 2005. Basionym: *Chondrorhyncha caudata* J. D. Ackerman, *Selbyana* 5: 299. 1981. Type: Ecuador, without locality, flowered in cultivation at SEL greenhouses, 17 June 1977, *J. D. Ackerman 1259* (holotype: SEL).

SYNONYMS

Stenia caudata (J. D. Ackerman) Dodson & D. E. Bennett, *Icon. Pl. Trop.*, ser. 2, 2: plate 181. 1989.

Ackermania caudata (J. D. Ackerman) Dodson & R. Escobar, *Orquideología* 18(3): 206. 1993, type species of *Ackermania*.

DESCRIPTION The sepals, petals, and lip are cream-white often with a bright yellow apex (alternatively white lightly suffused with yellow or green at apexes); the lip is white with a yellow apex, spotted red and suffused red along the margins or immaculate; the column is white with red-brown spots. The flower is resupinate. The lip is saccate, three-lobed, caudate at the apex, the tail points upwards and out. The callus is low, midline, and flat, forming a shallow ridge at mid length of the lip with a four-lobed apex. The column is shorter than its column foot, without wings, and not pubescent. PLATE 2.

COMMENT *Benzingia caudata* is distinguished from *B. palorae* by the lack of reflexing on the apical end of the lip midlobe and by differences in the ratio of the length of the column foot to the length of the rest of the column. The column foot of *B. caudata* is twice as long as the rest of the column, whereas in *B. palorae* the column foot is the same length as the rest of the column. Dodson and Bennett (1989) reported *Stenia caudata* as being from Peru, but this report is correctly *B. jarae*. The name combination remains valid; just the specimen used as an example was incorrect.

MEASUREMENTS Leaves 9–20 cm long, 1.5–3.5 cm wide; inflorescence 3–4 cm long; dorsal sepal 1.4–1.8 cm long, 0.5–0.6 cm wide; lateral sepals 1.8–2.1 cm long, 0.5–0.9 cm wide; petals 1.5–1.6 cm long, 0.6 cm wide; lip 1.3–1.6 cm long; column 1.0–1.1 cm long.

ETYMOLOGY Latin *caudatus*, ending with a tail-like appendage, describing the lip.

DISTRIBUTION AND HABITAT Known only from Ecuador, at 1000–1200 m in elevation.

FLOWERING TIME Not recorded.

Benzingia cornuta (Garay) Dressler,

Lankesteriana 5(2): 93. 2005. Basionym: *Chondrorhyncha cornuta* Garay, *Orquideología* 5: 20. 1970. Type: Colombia, Antioquia, *Gilberto Escobar 593* (holotype: AMES).

SYNONYM
Ackermania cornuta Garay, *Orquideología* 18: 206. 1993; photograph published in 2: 237. 2000.

DESCRIPTION The flower is white, the inside of the lip is yellow with red spots. The flower is resupinate. The dorsal sepal and petals cover the column, the lateral sepals embrace the lip from below. The lip is three- to five-lobed, the lateral four lobes are obscure and erose, the medial lip is concave, tubular, the apical margins lightly revolute. The lip apex has a prominent porrect, caudate point. The callus is a basal thickening with a medial keel or thickening, becoming longer apically as it spreads to the lateral lip edges (apical edge U-shaped). The column is widest at the stigma, slightly curved apically, does not hood over the pollinia, and is ventrally pubescent.

MEASUREMENTS Leaves 25 cm long, 2 cm wide; inflorescence 5 cm long; dorsal sepal 1.8 cm long, 0.5 cm wide; lateral sepals 2.5 cm long, 0.8 cm wide; petal measurements not given, about the same as dorsal sepal; lip 2.5 cm long, 2 cm wide; column 1 cm long.

ETYMOLGY Latin *cornutus*, horned, referring to caudate point on the lip.

DISTRIBUTION AND HABITAT Known from Colombia and Ecuador, at 1500 m elevation.

FLOWERING TIME Not recorded.

Benzingia estradae (Dodson) Dodson, *Icon.*

Pl. Trop., ser. 2, 5: plate 406, 1989; *Lindleyana* 10: 74. 1995. Basionym: *Chondrorhyncha estradae* Dodson, *Icon. Pl. Trop.* 1: plate 22. 1980, as "*Chondroryncha.*" Type: Ecuador, El Oro, 800–1100 m, 10 December 1979, *Dodson 9246* (holotype: SEL).

DESCRIPTION The sepals and petals are yellow-white; the lip is orange with red spots. The resupinate flowers are pendent or nodding. The lip is mildly saccate at the base, the apical margin is entire. The callus is a broad bilobed flap in the center of the saccate portion of the lip. The column is pubescent at the base, and at a right angle to the column foot. PLATE 3.

MEASUREMENTS Leaves 18 cm long, 3 cm wide; inflorescence 7 cm long; sepals 2.5 cm long, 0.9 cm wide; petals 2.5 cm long, 1.5 cm wide; lip 2.5 cm long, 1.5 cm wide; column 1.8 cm long.

ETYMOLOGY Latin *estriatus*, not striate, referring to the pattern of the orange spots on the lip.

DISTRIBUTION AND HABITAT Known only from Ecuador, at 800–1100 m in elevation, in extremely wet montane cloud forest.

FLOWERING TIME Most of the year.

Benzingia hajekii (D. E. Bennett &

Christenson) Dressler, *Lankesteriana* 5(2): 93. 2005. Basionym: *Ackermania hajekii* D. E. Bennett & Christenson, *Icon. Orch. Peruv.*, plate 602. 2001. Type: Peru, Pasco, Oxapampa, 1200 m, hort. Carlos Hajek, 9 March 1999, *Bennett 7890* (holotype: MOL).

DESCRIPTION The sepals and petals are pale yellow with darker apexes; the lip is orange-yellow with pale garnet-red-brown spots and short transverse bars. The column is pale yellow with the ventral surface and the foot of the column having brown spots. The anther is pale yellow. The resupinate flower is bell-shaped with slightly reflexed lateral sepals. The lip is obscurely five-lobed, suborbicular, obtuse, and concave with incurved suberect lateral margins. The apical lip margin has a blunt, obtuse point. The callus is transverse and bidentate with rounded teeth. The column is clavate, pubescent, and hirsute on the ventral surface with obtuse wings or widenings lateral to the stigma. PLATE 4.

MEASUREMENTS Leaves 34 cm long, 3.3 cm wide; inflorescence 4 cm long; dorsal sepal 1.5

cm long, 0.6 cm wide; lateral sepals 2.2 cm long, 0.6 cm wide; petals 2 cm long, 0.7 cm wide; lip 1.8 cm long, 1.7 cm wide; column 1.5 cm long, 0.4 cm wide.

ETYMOLOGY Named after Carlos Hajek of Peru, who collected and flowered the type plant.

DISTRIBUTION AND HABITAT Known only to Peru, at 1200 m in elevation, in wet montane forest.

FLOWERING TIME March.

Benzingia hirtzii Dodson, *Icon. Pl. Trop.*, ser. 2, 5: plate 406. 1989; *Lindleyana* 10(2): 74. 1995. Type: Ecuador, Imbabura, Santa Rosa de Chaco, 1150 m, 19 January 1987, *Dodson, Hirtz, Benzing & C. Luer 16893* (holotype: RPSC); type species of *Benzingia*.

SYNONYM
Benzingia litense Dodson & "Chase" ex Dodson, *Nat. Ecuador. Orch.* 1: 60. 1994, *nom. nud., de facto pro syn.*

DESCRIPTION The flowers are yellow; the sepals and petals are spotted with red-brown. The lip is yellow with an orange callus (sometimes the base of the lip and the callus are covered with red blotches) and abundantly spotted with red-brown. The flower is nonresupinate. The sepals are free and spreading. The lip is shallowly concave, scoop-shaped, and rotund, not surrounding the column, the apical margin reflexing mildly. The callus is flat, lying in the midline of the lip. The column is arcuate, slightly winged on each side of the stigma. The column foot is 0.8 cm long. PLATE 5.

MEASUREMENTS Leaves 12 cm long, 1.6 cm wide; inflorescence 2 cm long; dorsal sepal 1.7 cm long, 0.5 cm wide; lateral sepals 1.8 cm long, 0.5 cm wide; petals 1.7 cm long, 0.5 cm wide; lip 2.2 cm long, 2.2 cm wide; column 1.5 cm long.

ETYMOLOGY Named after Alexander Hirtz, who participated in the collection of this species.

DISTRIBUTION AND HABITAT Known only to northwestern Ecuador, from 725 to 1200 m in elevation.

FLOWERING TIME January to May.

Benzingia jarae (D. E. Bennett & Christenson) Dressler, *Lankesteriana* 5(2): 93. 2005. Basionym: *Ackermania jarae* D. E. Bennett & Christenson, *Brittonia* 47: 182. 1995. Type: Peru, Huánuco, Leoncio Prado, 1200 m, April 1994, *E. Jara P. ex Bennett 6578* (holotype: NY).

DESCRIPTION The sepals and petals are green-yellow to cream-yellow. The lip is orange-yellow with red-brown spots on interior surface; the apex of the lip is pale yellow, the column cream-yellow to yellow-orange, the column foot has red-purple markings, and the anther is white with brown markings. The flower is resupinate. The lip is obscurely three-lobed, saccate, the lateral lobes are erect, partially enclosing the column, and the midlobe has a recurved apicular tooth and a small falcate tooth laterally to each side. The callus is a transverse ridge below the midlobe of the lip. The column is straight, slightly dilated below the apex lateral to the stigmatic cavity. PLATE 6.

COMMENT *Benzingia jarae* is distinguished from *B. caudata* by its cream-yellow flowers with a deeper yellow almost orange lip, teeth on the lip edge lateral to the apicule, and an apicule less than half the length of the apicule of *B. caudata*. *Benzingia jarae* is the most saccate species in genus *Benzingia*. The Internet has several mislabeled pictures of *B. caudata* that most likely are of *B. jarae*.

MEASUREMENTS Leaves 10–15 cm long, 2.0–4.5 cm wide; inflorescence 4 cm long; dorsal sepal 1.5 cm long, 0.9 cm wide; lateral sepals 1.8 cm long, 1.3 cm wide; petals 1.5 cm long, 1 cm wide; lip 1.25 cm long, 0.75 cm wide; column 0.9 cm long, 0.4 cm wide.

ETYMOLOGY Named for Enrique Jara P., who collected the species.

DISTRIBUTION AND HABITAT Known only to Peru, at 1000–1400 m in elevation, in wet montane forest.

FLOWERING TIME March to April and October to November.

Benzingia palorae (Dodson & Hirtz)

Dressler, *Lankesteriana* 5(2): 93. 2005. Basionym: *Stenia palorae* Dodson & Hirtz, *Icon. Pl. Trop.*, ser. 2, 6: plate 583. 1989. Type: Ecuador, Pastaza, Río Palora, 800 m, 15 July to 15 October 1983, *Hirtz 1545* (holotype: SEL).

SYNONYMS
Chondrorhyncha palorae (Dodson & Hirtz) Senghas & G. Gerlach, *Orchideen* (Schlechter) 1/B(26): 1631. 1992.
Ackermania palorae (Dodson & Hirtz) Dodson & R. Escobar, *Orquideología* 18(3): 206. 1993.

DESCRIPTION The sepals and petals are white. The lip exterior is white with yellow borders and tip; the lip interior has brown markings. The flowers are resupinate. The lip is obscurely three-lobed and the lateral lobe edges embrace the column foot so that the lip appears saccate. The apical lip lobe is elongate, linear, and recurved along the margin of the lip. The callus is low, midline, and flat, forming a shallow ridge at mid length of the lip with a four-lobed apex. The column is expanded on each side of the slitlike stigma. PLATE 7.

COMMENT *Benzingia caudata* is distinguished from *B. palorae* by the lack of reflexing on the apical end of the lip midlobe. The column foot of *B. caudata* is twice as long as the rest of the column, whereas *B. palorae* has these two portions in equal size.

MEASUREMENTS Leaves 20 cm long, 4 cm wide; inflorescence 8 cm long; dorsal sepal 2 cm long, 1 cm wide; lateral sepals 2 cm long, 1.2 cm wide; petals 3 cm long, 1 cm wide; lip 0.6 cm long, 1.5 cm wide; column 2 cm long.

ETYMOLOGY Named for the location where the species was first found.

DISTRIBUTION AND HABITAT Known from Colombia, Ecuador, and Peru, at 800–1200 m in elevation, in very wet montane forest in deep shade.

FLOWERING TIME July to October.

Benzingia reichenbachiana (Schlechter)

Dressler, *Lankesteriana* 5(2): 93. 2005. Basionym: *Chondrorhyncha reichenbachiana* Schlechter, *Repert. Spec. Nov. Regni Veg.* 17: 15. 1921. Type: Costa Rica, *Endrés 557* (holotype: W).

DESCRIPTION The sepals and petals are cream-yellow to white. The lip is yellow with a pink-red suffusion in the apical thickening and sometimes the suffusion continues across the lip apically from the callus; the margins of the lip are clear of the suffusion. The flower is resupinate. The lateral sepals are reflexed with internal margins strongly involute and doubled to the mid length. The lip is concave, notched at the apex, said to be gullet-shaped, and reflexed at the apex mildly. The callus is platelike, with a dentate or laciniate margins, and narrowly bilobed (or just flat with a notch in the middle). The pollinia have a prominent stipe. PLATE 8.

COMMENT *Benzingia reichenbachiana* has many traits that morphologically make the species difficult to place in a particular genus. The lateral sepals roll to form a false spur, similar to other species of the *Chondrorhyncha* complex. The callus and lip are similar to those of *Euryblema*. These characteristics differ from those of other *Benzingia* species and suggest that *B. reichenbachiana* uses nectar deceit for pollination instead of fragrance attraction as do other *Benzingia* species. The stipe, viscidium, and column do fit within *Benzingia*, and actually, when one compares the column of *B. caudata* with the column of *B. reichenbachiana*, it is remarkable how similar they are both in shape and in coloration. Molecular data confirm this species belongs within *Benzingia*. It is the only species of *Benzingia* to be found in Central America.

MEASUREMENTS Leaves 15–30 cm long, 1.5–2.4 cm wide; inflorescence 1.2–1.5 cm long; dorsal sepal 1.3–1.5 cm long, 0.4–0.6 cm wide; laterals sepals 1.8–2.0 cm long, 0.4–0.6 cm wide; petals 1.8–2.2 cm long, 0.7–0.9 cm wide; lip 1.8–2.2 cm long, 1.6–1.8 cm wide; column 1 cm long.

ETYMOLOGY Named for Heinrich Gustav Reichenbach, who at his time was the authority of orchid taxonomy in Europe.

DISTRIBUTION AND HABITAT Known from Costa Rica, Panama, and Colombia, at 500–1400 m in elevation, in pre-montane rain forest.

FLOWERING TIME Throughout the year.

Benzingia thienii (Dodson) P. A. Harding, *comb. nov.* Basionym: *Cochleanthes thienii* Dodson, *Icon. Pl. Trop.* 1: plate 26. 1980. Type: Ecuador, Pichincha, Dos Ríos, 1200 m, 10 February 1963, *Dodson & Thien 2201a* (holotype: SEL).

SYNONYM
Chondrorhyncha thienii (Dodson) Dodson, *Selbyana* 7: 354. 1984.

DESCRIPTION The sepals and petals are yellow, and the lip is pale orange with red blotches at the base and over the callus. The column is white, yellow at the base with red spots or blotches at its ventral base. The dorsal sepal is variably recurved at the apex or concave and covering the column. The lateral sepals are reflexed back and up. The petals clasp the column laterally and are recurved at the apex. The lip is obovate, the lateral margins incurve over the column, the lip apex flares minimally but remains inrolled apically, and the apical lip margin is irregularly and shallowly lacerate. The thick callus is a transverse plate to about two-thirds the lip length, the apical margin is four-toothed with a midline keel. The column is widest at the stigma with minimal wings and a ventral keel that starts about mid column length and goes to the base. The viscidium is broad and shield- or trowel-shaped; the stipe is minimal.

COMMENT *Benzingia thienii* could fit into other genera depending on which characteristics are considered important. I feel that the thick callus extending across the entire width of the lip with a central keel, the column shape, and the ratio of the column foot to the column length place this species in *Benzingia* over other genera. Aspects of the plant that would exclude it from *Benzingia* are the green foliage, not gray-green as other *Benzingia* species, and the upright growth habit. Whitten et al. (2005) believe *Chondrorhyncha thienii* could belong to the genus *Euryblema*, but do not formally place it there. *Benzingia thienii* flowers are half the size of the flowers of the two species of *Euryblema*, and the callus is markedly thick apically, whereas the callus of *Euryblema* is a flat plate. The two also differ in column, column foot, and stipe shape. Further study may determine that *Benzingia thienii* deserves its own genus.

MEASUREMENTS Leaves 10–25 cm long, 2.0–4.5 cm wide; inflorescence 9 cm long; dorsal sepal 1.6 cm long, 0.9 cm wide; lateral sepals 2.5 cm long, 1 cm wide; petals 1.6 cm long, 1 cm wide; lip 2.5 cm long, 2 cm wide; column 3 cm long, 1.5 cm wide.

ETYMOLOGY Named for Leonard Thien, who participated in the discovery of the species.

DISTRIBUTION AND HABITAT Known from western Ecuador, at 1000–1550 m in elevation.

FLOWERING TIME February and March.

• •

The genus **Ackermania** was established by the publication of *Ackermania* Dodson & R. Escobar in *Orquideología* (1993, 18: 202). The saccate, tubular-lipped *Chondrorhyncha* species have been moved repeatedly about over the years. Some were moved to the genus *Stenia* by Calaway Dodson in 1989. In 1991, Karl Senghas established Section *Stenoides* Senghas & Gerlach in *Chondrorhyncha* and transferred some species of *Stenia* and *Chondrorhyncha* to that group. Dodson and Escobar felt Senghas' section was a natural group and used it form the genus *Ackermania*, dedicated to James Ackerman, an orchid taxonomist. The name, however, is a homonym of *Ackermannia*, a fungus, described in 1902, and therefore illegitimate.

The genus *Ackermania* was described as characterized by plants with a short rhizome lacking pseudobulbs with leaves that have sheaths, and by pendent foliage arranged in the shape of a fan. The leaves are veined. The single resupinate-flowered inflorescences emerge from leaf sheaths at the base. The flower lip is deeply tubular-sacciform, with a truncate, apiculate apex.

The species included in *Ackermania* originate from Colombia, Ecuador, and Peru, at 1000–1500 meters in elevation, on tree trunks in humid shady rain forests. All of the species—*A. caudata* (type species of the genus), *A. cornuta*, *A. hajekii*, *A. jarae*, and *A. palorae*—have been listed above as members of genus *Benzingia*, achieving what some molecular taxonomists say they are striving to achieve, less genera, and making the name *Ackermania*, and its illegitimacy, not a concern.

Chaubardia Reichenbach f., *Bot. Zeit. (Berlin)* 10: 671. 1852.

Heinrich Gustav Reichenbach originally proposed the genus *Chaubardia* in 1852, with the type species being *C. surinamensis*, recognizing that this group of species did not belong in the genus *Zygopetalum* or in the genus *Huntleya*. Based on a poorly preserved specimen with a single flower and plant, he described the genus as having plants without a pseudobulb. In his written text Reichenbach states that the plants are like *Kefersteinia* in growth habit. *Chaubardia* remained without other specimens until G. C. K. Dunsterville found a plant in the wild, clearly with a small pseudobulb, and since then several specimens have been found in herbaria, all with pseudobulbs. Eric Christenson (pers. comm.) reports that the type specimen does indeed have a pseudobulb, which was not appreciated at the time of first description.

Chaubardia species have very small inconspicuous pseudobulbs at the base, but otherwise the vegetative growths are fan-shaped. The species possess open, flat flowers borne on more or less upright inflorescences that don't extend past the foliage. The inflorescence has one to three bracts along its length. The sepals and petals of *Chaubardia* are narrower than those of *Huntleya*, producing a thin star-shaped flower. The lateral sepals are distinctly gibbose at the base. The lip is rhomboid, unguiculate, and continuous with the column foot. The long lip base kinks or reflexes backward and upward. The lip has two portions, the epichil (the midsection or apical section) and the hypochil (the basal section). The callus is conspicuously long-toothed, in some species becoming fimbriate. Like the columns of *Huntleya*, the columns of *Chaubardia* flowers have lateral wings and a hooded clinandrium; however, the anther is not crested and the stigma is ovate.

The genus *Chaubardia* is found from Trinidad and Venezuela, across South America, and south to Bolivia and Peru throughout the Amazon Basin, from sea level to 1200 meters in elevation, in humid forests.

The molecular data strongly support separating *Chaubardia* from *Huntleya* and the monophyletic nature of this group. Molecular studies were not used to include *Hoehneella* in the genus *Chaubardia* because of the lack of material. *Hoehneella* species differ by the lack of basal reflexing of the lip and by merely raised callus teeth; *Chaubardia* species have basal reflexing of the lip and long, fimbriate callus teeth. The two genera share the lack of a minimal stipe, ovate stigma, and hooded clinandrium.

Etymology
Named for Lois Athanase Chaubard, a French botanist who worked on the *Flora of Europe* and was a friend of Reichenbach.

List of the Species of *Chaubardia*
*Chaubardia heteroclita**
*Chaubardia klugii**
*Chaubardia surinamensis** (type species of *Chaubardia*)
* Molecular sampling confirms placement of this species within the genus *Chaubardia*.

Key to the Species of *Chaubardia*
1a. Callus teeth rounded, lip and callus blue . . . . *C. surinamensis*
1b. Callus teeth fimbriate (thin and pointed) . . . . go to 2
2a. Flower white, pollinia yellow, lip pubescent . . . . *C. klugii*
2b. Flower multicolored, with sepals and petals yellow to brown, lip white to pale purple, callus white to blue . . . . *C. heteroclita*

Chaubardia heteroclita (Poeppig & Endlicher) Dodson & D. E. Bennett, *Icon. Pl. Trop.*, ser. 2, 1: plate 23. 1989. Basionym: *Maxillaria heteroclita* Poeppig & Endlicher, *Nov. Gen. Sp. Pl.* 1: 37. 1836. Type: Lectotype (Christenson 2007). Original drawings of a flower and floral dissections, *Herb. Reichenbach 51473* (lectotype: W).

SYNONYMS
Zygopetalum rhombilabium C. Schweinfurth, *Amer. Orch. Soc. Bull.* 7: 422. 1944.
Warczewiczella heteroclita (Poeppig & Endlicher) Hoehne, *Arq. Bot. Estad. São Paulo*, n.s., f.m., 2: 126. 1952, as *Warscewiczella*.
Cochleanthes heteroclita (Poeppig & Endlicher) R. E. Schultes & Garay, *Bot. Mus. Leafl.* 18: 325. 1959.
Cochleanthes rhombilabia (C. Schweinfurth) Senghas, *Orchidee (Hamburg)* 17: 192. 1966.
Huntleya heteroclita (Poeppig & Endlicher) Garay, *Orquideología* 4: 146. 1969.

Chondrorhyncha rhombilabium (C. Schweinfurth) Fowlie, *Orch. Digest* 48: 225. 1984.

DESCRIPTION The sepals and petals are pale to dark tan-yellow with four to seven wide brick-red veins and a lip which is white to pale purple (there is a gold-lipped form from Peru); the callus is white to darker blue-purple, the column is white to lilac-purple, the anther is cream, and the pollinia are yellow. The lip could be thought to have three lobes, but the lateral lobes are just minimal flaps. The hypochil is concave (the least concave of the genus), with 12 keels or lamellae. The lamellae of the hypochil become the free fimbriae of the callus, and the apexes of the lamellae are round or obtuse. The column is arcuate with broad triangular quadrate wings at the mid length and a large central keel on the ventral surface at the base. The stigma is elliptic and is less broad than in other species in the genus. PLATE 9.

MEASUREMENTS Leaves 15–55 cm long, 3.0–6.5 cm wide; inflorescence 15 cm long; dorsal sepal 2.5 cm long, 0.9 cm wide; lateral sepals 3.3 cm long, 0.7 cm wide; petals 2.7 cm long, 0.7 cm wide; lip 3.9 cm long, 1.6 cm wide; column 1.1 cm long.

ETYMOLOGY Greek *hetero*, different or uneven, and *clete*, wedge, referring to the different teeth of the callus.

DISTRIBUTION AND HABITAT Known from Bolivia, Brazil, Ecuador, and Peru, at 700–1500 m in elevation, in montane wet forest.

FLOWERING TIME November to March.

Chaubardia klugii (C. Schweinfurth)

Garay, *Orquideología* 8: 34. 1973. Basionym: *Zygopetalum klugii* C. Schweinfurth, *Bot. Mus. Leafl.* 15: 159. 1952. Type: Peru, Loreto, near Iquitos, 100 m, July 1937, *G. Klug 10109* (holotype: AMES).

SYNONYM
Cochleanthes klugii (C. Schweinfurth) R. E. Schultes & Garay, *Bot. Mus. Leafl.* 18: 325. 1959.

DESCRIPTION The flowers are white, with the exception of the stipe, which is pale pink, and the pollinia, which are pale yellow. The lip is pubescent overall, including the teeth of the hypochil and callus. The hypochil is concave and cup-shaped, the epichil is elliptic, cuneate, and small compared to the hypochil. The apex of the lip is pointed. The column has wings at mid length. The stigma is broad and obovate-orbicular. The stipe and viscidium are convex and pointed. PLATE 10.

MEASUREMENTS Of a Peruvian plant (Dodson and Bennett 1989: plate 24): Leaves 5–15 cm long, 1.5–3.0 cm wide; inflorescence 5 cm long; dorsal sepal 3 cm long, 0.7 cm wide; lateral sepals 2.4 cm long, 0.8 cm wide; petals 2.4 cm long, 0.5 cm wide; lip 2.3 cm long, 0.6 cm wide; column 1.8 cm long, 0.7 cm wide.

Of a Bolivian plant (Dodson and Vásquez 1989: 312): Leaves 25 cm long, 5 cm wide; dorsal sepal 1.7 cm long, 0.6 cm wide; lateral sepals 2 cm long, 0.7 cm wide; lip 1.3 cm long, 0.7 cm wide.

ETYMOLOGY Named for Guillermo Klug, who collected plant specimens throughout the tropics.

DISTRIBUTION AND HABITAT Known from Colombia, Bolivia, Ecuador, and Peru, at 100–1700 m in elevation, in tropical rain forest.

FLOWERING TIME June to September.

Chaubardia surinamensis Reichenbach

f., *Bot. Zeit. (Berlin)* 10: 672. 1852. Type: Suriname (holotype: W); type species of *Chaubardia*.

SYNONYMS
Zygopetalum trinitatis Ames, *Sched. Orchid.* 3: 21. 1923. Type: Trinidad, Rio Claro, 7th mile Guyaguayare Road, 5 December 1922, *R. A. Farfan s.n.* (holotype: NY).
Hoehneella santos-nevesii Ruschi, *Publ. Arq. Publico Estad. Esp. Santo*: 3. 1945.
Cochleanthes trinitatis (Ames) R. E. Schultes & Garay, *Bot. Mus. Leafl.* 18: 326. 1959.
Hoehneella trinitatis (Ames) Fowlie, *Orch. Digest* 33: 231. 1969.

DESCRIPTION The sepals and petals are white, the lip is blue-white to white, and the callus is blue or pink, while some plants have blue only along the upper face of the lip crest. The lip is ovate and hirsute on the apical half, the hypochil of the lip has a central low broad rib which is continuous with the claw of the lip. The callus has nine large lamellae or keels, and extends more than a third of the lip length. The column is curved with a pair of auricle-like wings on each side of the stigma, and is pubescent ventrally below the stigma. The stigma is broad and obovate-orbicular. The viscidium is rhombic, with a short keel in the center separating the two pairs of pollinia.

COMMENT *Cochleanthes trinitatis* has only seven

lamellae in the callus and is found only on Trinidad.

MEASUREMENTS Leaves 7.5–13.0 cm long, 1.5 cm wide; inflorescence 5 cm long; dorsal sepal 1.2 cm long, 0.5 cm wide; lateral sepals 1.1 cm long, 0.4 cm wide; petals 1 cm long, 0.3 cm wide; lip 0.7 cm long, 0.6 cm wide; column 0.8 cm long.

ETYMOLOGY Named for Suriname, where the type specimen was found.

DISTRIBUTION AND HABITAT Known from all tropical South America extending south to Bolivia and Peru, at 250–500 m in elevation, in tropical, hot, wet forests.

FLOWERING TIME Most of the year.

Chaubardiella Garay, *Orquideología* 4: 146. 1969.

The genus *Chaubardiella* was established in 1969 by Leslie Garay who included five species: *C. calceolaris*, *C. chasmatochila*, *C. saccata*, *C. subquadrata*, and *C. tigrina*. The species have in common the shape of the short thick column and a long filiform rostellum, in addition to a cup-shaped lip which articulates with the short foot of the column. Since 1969, additional species have been described while *C. calceolaris* and *C. saccata* have been removed to the genus *Stenia*.

Plants of the genus are caespitose epiphytes. The pseudobulbs are absent or reduced but there is a distinct short stem. The small fan-shaped plants are produced from imbricating foliaceous sheaths on the short stem. The deep green leaves, widest at mid length, taper to a sheath at the base. The one-flowered inflorescence is produced from the base of the sheath and angles downward. For plants in pots, this presentation as well as the large size of the flower causes the inflorescence to lie on the surface of the substrate. The inflorescence has at most one bract mid length and up to three bracts subtending the ovary. The flowers are fleshy and the sepals and petals are spreading. The flowers are often nonresupinate though this trait is variable within the species and I believe not diagnostic based on the pictures I have seen. The lateral sepals do not extend backward from the flower to form false nectar-tubes as in *Chondrorhyncha*. The lip is concave for most to all of its length, the lateral edges of the lip do not surround the column. The lip is articulated to the column foot. The basal callus is fleshy and platelike. The column is usually short and thick, and often has fleshy wings extending on either side. Some species have a clinandrium with an apical margin that is elevated and three-lobed; other species have just a simple apical clinandrium. The pollinia are in two unequal pairs attached to a quadrate, short to medium-sized stipe and a hooked viscidium. The rostellum is long, filiform, and hooked.

Chaubardiella species have developed a hooked rostellum perhaps to attach to the base of one of the legs of *Euglossine* bees (Whitten et al. 2005).

In nature these plants grow in dense shady forests on steep slopes from 300 to 1500 meters in elevation, in very wet environments.

Molecular data show the genus *Chaubardiella* to be a strongly supported group, and it is closely related to *Aetheorhyncha*, *Ixyophora*, *Pescatorea*, and *Warczewiczella*.

Etymology

Diminutive of *Chaubardia*.

List of the Species of *Chaubardiella*

Chaubardiella chasmatochila
Chaubardiella dalessandroi
Chaubardiella delcastillo
Chaubardiella hirtzii
*Chaubardiella pacuarensis**
*Chaubardiella pubescens**
Chaubardiella serrulata
*Chaubardiella subquadrata**
*Chaubardiella tigrina** (type species of *Chaubardiella*)

* Molecular sampling confirms placement of this species within the genus *Chaubardiella*.

Key to the Species of *Chaubardiella*

1a. Flower densely spotted all over with maroon, red, or brown . . . . go to 2
1b. Flower not densely spotted all over . . . . go to 6
2a. Callus with shallow basal keels that form a transverse plate, callus pubescent . . . . *C. pubescens*
2b. Callus composed of keels that end apically in a raised U-shaped ridge . . . . go to 3
3a. Callus with central keel that becomes a tooth apically . . . . *C. dalessandroi*
3b. Callus without central keel . . . . go to 4
4a. Lip with apical notch, sepals concave . . . . *C. delcastillo*
4b. Lip with apical point, sepals convex . . . . go to 5
5a. Lip obovate, widest at base and gradually comes to point . . . . *C. tigrina*
5b. Lip subquadrate, square, with one side of square folded to form a point . . . . *C. hirtzii*
6a. Lip margin serrulate . . . . *C. serrulata*
6b. Lip margin smooth or with rounded irregularities . . . . go to 7
7a. Lip narrows apically, lip with apical notch, flower white . . . . . *C. pacuarensis*
7b. Lip subquadrate . . . . go to 8
8a. Lip open, not inrolled, flower yellow . . . . *C. chasmatochila*
8b. Lip inrolled at margins, flower light yellow with pink suffusion . . . . *C. subquadrata*

Chaubardiella chasmatochila (Fowlie)

Garay, *Orquideología* 4: 148. 1969. Basionym: *Stenia chasmatochila* Fowlie, *Orch. Digest* 29: 347. 1965. Type: Costa Rica, Sarapiqui, La Laguna del Cerro Congo, 650 m, 14 km northwest of Cariblanco, March 1961, *Clarence Kl. Horich s.n.* (holotype: UCLA).

DESCRIPTION The usually nonresupinate flowers are buttercup-yellow, sometimes with minimal red spots or dots. The lip is subsessile, minimally clawed, concave, subquadrate, with a margin that is entire and cupped, not rolled inwardly. The callus is a semicircular plate or basal thickening without ornament. The column is fleshy with a short foot, and the clinandrium is acute. PLATE 11.

COMMENT Some taxonomists have listed *Chaubardiella chasmatochila* and *C. subquadrata* as synonyms. There are many differences between them including the size of the flowers, the shape of the callus, and the coloration of the flowers. The simple cupped lip of *C. chasmatochila* allows the callus and inner lip to be easily seen from the front, whereas the deeply cochleate lip of *C. subquadrata* keeps the callus and inner lip hidden from the front.

MEASUREMENTS Leaves 10–16 cm long, 1.8–3.5 cm wide; dorsal sepal 2.2 cm long, 1.1 cm wide; lateral sepals 2.3 cm long, 1 cm wide; petals 2 cm long, 1 cm wide; lip 1.9 cm long, 1.8 cm wide; column 0.6 cm long.

ETYMOLOGY Greek *chasma*, fissure or deep cleft, and *chilus*, lipped, referring to the lip shape.

DISTRIBUTION AND HABITAT Known from Costa Rica to Colombia, at 600–1500 m in elevation.

FLOWERING TIME March to October.

Chaubardiella dalessandroi Dodson &

Dalström, *Ic. Pl. Trop.*, ser. 1, 10: plate 911. 1984. Type: Ecuador, Morona-Santiago, 1500 m, 27 December 1982, *Dalström 429* (holotype: SEL).

DESCRIPTION The flowers are tan-brown, densely dotted with dark purple-brown, the column is dark wine-purple, the lip callus is dark purple on both ends, and the callus teeth have tan-purple dots in the center. The flowers are usually resupinate. The dorsal sepal and petals are spatulate. The lip is quadrate, obtuse at the apex with a central notch. The callus is a fan-shaped raised plate with 15 teeth on the apical margin. A prominent central keel in the callus runs the length of the callus, becoming a central tooth apically. The column is winged, and the clinandrium is acute. PLATE 12.

COMMENT The prominent central keel in the callus is fairly easy to see and separates this species from *Chaubardiella hirtzii* and the other heavily marked species.

MEASUREMENTS Leaves 5–12 cm long, 2–4 cm wide; inflorescence 3 cm long; dorsal sepal 2.0–2.5 cm long, 1.0–1.2 cm wide; lateral sepals 2.2–2.6 cm long, 1.0–1.4 cm wide; petals 2.4–3.3 cm long, 1.0–1.4 cm wide; lip 2 cm long, 1.6 cm wide; column 0.7 cm long and wide.

ETYMOLOGY Named for Dennis Dalessandro, for his contribution to the knowledge of the orchids of Ecuador.

DISTRIBUTION AND HABITAT Known only from Ecuador, at 1500–1800 m in elevation, in wet montane forest in deep shade.

FLOWERING TIME December.

Chaubardiella delcastillo D. E. Bennett

& Christenson, *Icon. Orch. Peruv.*, plate 423. 1998. Type: Peru, Department of Cuzco, Quillabamba, San Pantuari, 900 m, September 1995, *O. del Castillo ex Bennett 7309* (holotype: MOL).

DESCRIPTION The sepals and petals have a base color of green densely marked with red-brown spots and bars. The exterior surface has pale diffuse garnet-red markings, and the interior surface is a dull green-yellow densely transversed by ruddy brown spots and bars. The bars on the basal third of the tepals are broad on a more yellow base color. The lip exterior is the same color as the tepals. The lip interior has a base color of dark yellow that is only faintly visible between the barring, and is densely traversed by glossy dark red-brown stripes; the saccate base of the lip has very dark black-brown glossy blotches. The column is red, densely overlaid with very dark red spots and markings. The anther is pale cream-white and the viscidium is translucent, glossy, and off-white. The plant has four leaves per growth and the flowers are usually nonresupinate. The sepals are concave. The dorsal sepal is elliptic and the petals are minimally spatulate with rounded apexes. The lip is one-lobed, deeply concave, and emarginate. The callus has 10 or more radiating keels becoming a low, shallowly emarginate transverse flap. The column is short, stout, dilated, and broadly winged. The clinandrium apical margin is elevated and three-lobed.

COMMENT This species is very similar to *Chaubardiella hirtzii*, but has a simple callus, a lip that is less saccate, concave sepals, and a narrow nonreflexed or nonrecurved skirt on the lip.

MEASUREMENTS Leaves 11.5–18.5 cm long, 2.0–2.7 cm wide; dorsal sepal 1.8 cm long, 1 cm wide; lateral sepals 2 cm long, 1 cm wide; petals 1.8 cm long, 1 cm wide; lip 1.5 cm long, 1.4 cm wide; column 0.8 cm long, 0.5 cm wide.

ETYMOLOGY Named for Oliveros del Castillo, who collected the type plant.

DISTRIBUTION AND HABITAT Known only from southern Peru, at 900 m in elevation, in wet montane forest.

FLOWERING TIME March to April, and September.

Chaubardiella hirtzii Dodson, *Icon. Pl. Trop.*, ser. 2, 5: plate 412. 1989. Type: Ecuador, Napo, Río Yasupino, 500 m, October 1985, *Hirtz 1191* (holotype: SEL).

DESCRIPTION The flowers are white densely blotched with dark wine-red, and the anther cap is white. The flowers are usually resupinate. The dorsal sepal and petals are spatulate with acute apexes. The lip is ovate or subquadrate, and the apical point, formed by folds of the lip, deflexes down with prominent midline concavity. The base of the lip is truncate with a short claw. The callus is a raised horseshoe-shaped protuberance with indistinct ribs on the basal side. The column is excavate on the underside, and expands on each side of the stigma to form shallow wings. The clinandrium is acute. PLATE 13.

COMMENT *Chaubardiella hirtzii* is very similar to *C. tigrina* but has a subquadrate lip with a reflexed apical portion forming the central concavity, whereas *C. tigrina* has an obovate lip with a definite point or tail of tissue. The callus of both species is horseshoe-shaped with no central keel, but keels of *C. hirtzii* are less raised. Compared to *C. tigrina*, *C. hirtzii* has smaller or absent auricles or wings on the column.

MEASUREMENTS Leaves 10–15 cm long, 2.5 cm wide; inflorescence 4 cm long; dorsal sepal 1.5–2.6 cm long, 0.9–1.2 cm wide; lateral sepals 1.5–2.7 cm long, 0.8–1.2 cm wide; petals 1.6–2.5 cm long, 1.1–1.2 cm wide; lip 1.8 cm long, 2 cm wide; column 0.8–1.0 cm long.

ETYMOLOGY Named for Alexander Hirtz, who collected the type plant.

DISTRIBUTION AND HABITAT Known only from eastern Ecuador and Peru, at 480–550 m in elevation, on tree trunks in deep shade in tropical wet forest.

FLOWERING TIME Sporadically most of the year.

Chaubardiella pacuarensis Jenny, *Orchidee (Hamburg)* 40(3): 91. 1989. Type: Costa Rica, Río Pacuare, 1200 m, 1986, *Jenny 3* (holotype: G).

DESCRIPTION The translucent milk white flowers are usually nonresupinate. The petals are spatulate-obovate (widening at mid length, widest at three-quarters length). The lip is cochleate with a central depression which originates from the callus, the lip margins are smooth, and the apex is emarginate. The callus is semicircular on the apical margin and has smooth margins. The column is winged, and the clinandrium is acute. PLATE 14.

MEASUREMENTS Leaves 8–10 cm long, 1.6–2.2 cm wide; inflorescence 4–5 cm long; sepals 2.2 cm long, 1.2 cm wide; petals 2.2 cm long, 1.2 cm wide; lip 1.8–2.0 cm long and wide; column 0.9 cm long.

ETYMOLOGY Named for the location where the species was first collected.

DISTRIBUTION AND HABITAT Known from Costa Rica and Panama, at 1000–1400 m in elevation, in wet tropical forest.

FLOWERING TIME September to October.

Chaubardiella pubescens J. D. Ackerman, *Selbyana* 5: 297. 1981. Type: Colombia, Chocó, vicinity of San Juan River, flowered in cultivation at SEL, 15 May 1979, *J. D. Ackerman 1379* (holotype: SEL).

DESCRIPTION The flowers are dark yellow and variously spotted maroon, and the lip has abundant spots that increase basally so that most of the inside appears maroon. The flowers are usually nonresupinate. The sepals are pubescent at the base, the petals are glabrous, and both sepals and petals are 7- to 10-nerved. The lip margin is minimally erose. The callus is a thick, fleshy, transverse plate whose apex is barely free with a sinuate margin; the callus is slightly pubescent at the front edge. The column is winged and densely pubescent especially on the underside, except for the anther, rostellum, and stigma. The column apex is flat without a point. PLATE 15.

COMMENT *Chaubardiella pubescens* is the only species in the genus with a column apex that is flat without a point. The species is said to have the largest flower in the genus.

MEASUREMENTS Leaves 8–18 cm long, 3–4 cm wide; inflorescence 1–3 cm long; dorsal sepal 2.5 cm long, 1 cm wide; lateral sepals 2.2 cm long, 0.9–1.0 cm wide; petals 2.1–2.2 cm long, 0.9–1.0 cm wide; lip 1.4 cm long, 1.2–1.3 cm wide; column 1.1–1.2 cm long.

ETYMOLOGY Latin *pubescens*, downy with short hairs, referring to the surface of the flower.

DISTRIBUTION AND HABITAT Known only from Colombia, elevation not recorded.

FLOWERING TIME May in cultivation.

Chaubardiella serrulata D. E. Bennett & Christenson, *Ic. Orchid. Peruv.*, plate 424. 1998, as "*serrulatum*." Type: Peru, Department of Cuzco, Quillabamba, 909 m, 26 September 1995, *O. del Castillo ex Bennett 7307* (holotype: MOL).

DESCRIPTION The flowers are yellow-green with a pale yellow column, the anther is yellow-white, and the pollinia are pale yellow. The flowers are usually nonresupinate. The concave lip is unlobed, sessile, acute at the apex, with minutely serrulate margins. The callus is fleshy and is an inverted U-shaped ridge with low radiating ribs basally within the U. The column is extremely fleshy with short prominent wings. The clinandrium is acute.

MEASUREMENTS Leaves 15–17 cm long, 2.8–3.0 cm wide; inflorescence 4.5 cm long; sepals 1.5–1.7 cm long, 0.8–0.9 cm wide; petals 1.5–1.7 cm long, 0.8–0.9 cm wide; lip 1.4 cm long, 1.8 cm wide; column 0.4 cm long.

ETYMOLOGY Latin *serratus*, saw-edged with teeth pointing forward, and *ul*, diminutive, referring to the margin of the lip.

DISTRIBUTION AND HABITAT Known only from southern Peru, at 900 m in elevation, in wet montane forest.

FLOWERING TIME September, recorded as June in cultivation.

Chaubardiella subquadrata (Schlechter)

Garay, *Orquideología* 4: 149. 1969. Basionym: *Kefersteinia subquadrata* Schlechter, *Repert. Spec. Nov. Regni Veg. Beih.* 19: 300. 1923. Type: Costa Rica, San Ramon, 1921, *G. Acosta s.n.* (holotype: B, destroyed; lectotype: AMES).

SYNONYM
Chondrorhyncha subquadrata (Schlechter) L. O. Williams, *Ceiba* 5: 195. 1956.

DESCRIPTION The flowers are cream-yellow with a light pink suffusion. The flowers are usually nonresupinate. The concave lip is subsessile, clawed, subquadrate, with a margin that is irregularly subcrenulate and inrolled. The callus is a semicircular plate or basal thickening without ornament. The column is fleshy, slightly decurved, with a short foot, and the clinandrium is acute. PLATE 16.

COMMENT Several photos of plants named *Chaubardiella subquadrata* were sent to me, all of which show plants that I believe to be yellow forms of *C. tigrina*. The photograph of a supposed *C. subquadrata* in *Native Colombian Orchids* (R. Escobar 1994–1998, 1: 77) also shows, I think, a yellow form of *C. tigrina* based on the pointed lip. The type is based on a drawing by Rudolf Schlechter that clearly shows a notched square lip. Photographs labeled *C. subquadrata* in Garay (1969: 149) match the type drawing in lip shape. Garay says he has seen material from Colombia, Ecuador, and Costa Rica that fits this type drawing. A key in his article separates the two species: *C. tigrina* is "white with numerous transverse maroon bars; lip saccate with a reflexed acute apex; crest transverse triangular, fleshy plate above the base," while *C. subquadrata* is "cream colored with a light pinkish suffusion; lip cochleate, subquadrate when expanded, bilobed at obtuse apex; crest a semicircular bilobed fleshy plate."

Plate 16 in the present volume pictures what I believe is a yellow form of *Chaubardiella tigrina*, sent as *C. subquadrata*. The lip is pointed and the segments are long, but the color is correct for *C. subquadrata*. A photograph of *C. subquadrata* in *Orquideología* 25(1): 107 shows the notched lip and shorter segments, but the color and the color patterns are very similar to those in Plate 16 of the present volume. Determining species from a photo is hard as one can't see the details of the callus. I include the plate as it could indeed be *C. subquadrata*, but it may be a separate species or form of another species also.

MEASUREMENTS Leaves 10–17 cm long, 0.8–1.5 cm wide; dorsal sepal 1.5 cm long, 0.7 wide; lateral sepals 1.6 cm long, 0.7 cm wide; petals 1.5 cm long, 0.7 cm wide; lip 1.3 cm long, 1.2 cm wide; column 0.7 cm long.

ETYMOLOGY Latin *sub-*, somewhat, and *quadratus*, square, referring to the shape of the lip.

DISTRIBUTION AND HABITAT Known from Costa Rica to Ecuador, at 1300 m in elevation.

FLOWERING TIME March to June.

Chaubardiella tigrina (Garay &

Dunsterville) Garay, *Orquideología* 4: 149. 1969. Basionym: *Chaubardia tigrina* Garay & Dunsterville, *Venez. Orchid. Ill.* 2: 72. 1961. Type: Venezuela, near the base of Angel Falls, 800 m, *Dunsterville 571* (holotype: AMES); type species of *Chaubardiella*.

DESCRIPTION The flower color is variable but generally tinted more or less yellow, brown, and lavender. The sepals and petals are pale white, yellow, or light lavender with transverse brown-purple mottling. The lip lines are pale brown-lavender to maroon-red with a base color of pale dark yellow or pale lavender, the column is white or purple with spots of dark rose-red, and the anther is cream to white. The flowers are usually nonresupinate. The broad lip is deeply concave with a retuse apex. The callus inner margin is horseshoe-shaped, with a thickened apex; the outer margins are retuse and free. The center of the lip callus has three low, short convergent ridges and other minor ridges laterally. The column is stout with prominent wings. The clinandrium margin is elevated and three-lobed. PLATE 17.

COMMENT *Chaubardiella tigrina* is very similar to *C. hirtzii* but has an obovate lip, whereas *C. hirtzii* has a subquadrate lip with an apical portion that includes a central concavity. The callus of both species is a horseshoe-shaped U with no central keel, but keels of *C. hirtzii* are less raised and the auricles or wings on the column are smaller or lacking compared to *C. tigrina*. I do think there may be two forms of *C. tigrina*, one with elliptic sepals and petals and another with spatulate sepals and petals. I was hoping one of these forms could have been determined to be one of the other species here. Without a closer look, I have to say both are *C. tigrina*, but perhaps there are other also differing features and the forms deserve separation.

MEASUREMENTS Leaves 19 cm long, 2.5 cm wide; inflorescence 3–5 cm long; dorsal sepal 1.7 cm long, 0.7 cm wide; lateral sepals 1.8 cm long, 0.8 cm wide; petals 1.8 cm long, 0.7 cm wide; lip 1.2 cm long, 0.8 cm wide; column 0.5 cm long.

ETYMOLOGY Latin *tigrinus*, tigerlike or barred like a tiger, referring to markings on the flower.

DISTRIBUTION AND HABITAT Known from Colombia, Venezuela, Ecuador, and Peru, at 400–800 m in elevation, found in tropical wet forest.

FLOWERING TIME January to April.

Chondrorhyncha Lindley, *Orch. Linden.*: 12. 1846.

John Lindley established the genus *Chondrorhyncha* in 1846, describing *C. rosea* as an epiphyte without pseudobulbs, having papery leaves and sepals that were obliquely inserted on the column. A century later Garay (1969) described the genus more thoroughly, characterizing it as having a three-part rostellum, of which the median tooth is commonly more developed than the lateral ones, a prominent column foot, and the characteristic position of the callus on the lip. Then in 2005, Whitten et al. narrowed the definition of the genus based on molecular and morphological characters to having a lip with a two-toothed callus that narrows apically and having an ovate viscidium without a distinct stipe attached to edge of viscidium. Whitten et al. state that the species are restricted to northern South America and list *C. caquetae*, *C. fosterae*, *C. hirtzii*, *C. macronyx*, *C. rosea*, *C. suarezii*, and *C. velastiguii* as being included in *Chondrorhyncha sensu stricto*.

To my eye, the two-toothed callus trait is seen in only a few of these species, and the callus is more of a plate with several central low keels with the centermost keel being prominent. The apical edge of the plate is free of the lip surface and the edge has two to several teeth that project apically. *Chondrorhyncha rosea* and *C. panguensis* are the only species in the list above with two teeth. If the callus is reinterpreted then one can add *C. inedita* to the above list. (I have left *C. macronyx* out, as I don't understand what that species truly is, and *C. fosterae* fits better with *Ixyophora*.)

Plants of the genus *Chondrorhyncha sensu stricto* are tufted epiphytic herbs without pseudobulbs. The leaves are distichous, though the stem is not always discernible. The leaves are acute or acuminate and contracted below into conduplicate petioles, and the leaves are generally wider than the leaves of other species that used to be considered within the genus *Chondrorhyncha*. *Chondrorhyncha velastiguii* and *C. suarezii* have elliptical leaves. The inflorescences are slender with one flower, less than half the length of the leaves, and are produced from axils of the lower leaves or bracts. The flowers range from small and inconspicuous to somewhat large and conspicuous. They can be found in green to yellow hues. The sepals are subequal; the apically concave dorsal sepal is held either erect or horizontal and covers the column. The lateral sepals are retrorse in many species and obliquely inserted on the short column foot or column base, and the lateral sepals inroll along their length forming tubes. The petals flex forward covering the column, are subequal to the dorsal sepal or broader, and are often widest at the apical quarter. The lip is obovate with its apex often notched, the lateral lobes or margins of the lip are erect. The lip has a minimal claw (a contraction at the base) on some species and is adnate to, or articulated with, the short column foot, sometimes forming a very short mentum or chin. The callus of the lip is broad or narrow depending on the species, being more or less fleshy and free at its denticulate apex. The column is semiterete, slender or broadly clavate apically, and sometimes narrowly winged. The column ventral surface may or may not have a keel at its base and the base of the column is produced into a short foot. The clinandrium is at an oblique 90-degree angle horizontal to the axis of the column, and the clinandrium is hooded. The short column foot is held 150–160 degrees from the main column. The four pollinia are unequal, waxy, with a minimal stipe, and have a triangular lanceolate viscidium.

Etymology
Greek *chondros*, cartilage, and *rhynchos*, beak, describing the rostellum.

List of the Species of *Chondrorhyncha*
*Chondrorhyncha caquetae**
*Chondrorhyncha hirtzii**
Chondrorhyncha inedita
Chondrorhyncha manzurii
Chondrorhyncha panguensis
*Chondrorhyncha rosea** (type species of *Chondrorhyncha*)
Chondrorhyncha suarezii
Chondrorhyncha velastigui
* Molecular sampling confirms placement of this species within the genus *Chondrorhyncha*.

Key to the Species of *Chondrorhyncha*
1a. Callus at lip mid length obviously raised from surface of lip . . . . go to 2
1b. Callus at lip mid length adpressed to lip surface . . . . go to 4
2a. No keels at base of lip, three-toothed callus at lip mid length . . . . . .*C. rosea*
2b. Keels of callus start at base of lip. . . . . go to 3
3a. Callus of three keels that become three teeth apically, flower green to creamy yellow, lip striped with longitudinal red-brown lines . . . .*C. caquetae*
3b. Callus of five keels that become three teeth apically, midlobe of lip with warty veins, tepals white, lip pink to purple . . . . *C. inedita*

4a. Callus with no medial tooth or, if present, medial tooth shorter than
lateral teeth . . . . go to 5
4b. Callus with medial tooth longer than lateral teeth . . . . go to 6
5a. Callus two-toothed. . . . . *C. panguensis*
5b. Callus four-toothed . . . . *C. suarezii*
6a. Dorsal sepal flat transversely and reflexed back at apex . . . . *C. manzurii*
6b. Dorsal sepal concave . . . . go to 7
7a. Lip with spots . . . . *C. velastigui*
7b. Lip without spots . . . . *C. hirtzii*

Chondrorhyncha caquetae Fowlie, *Orch. Digest* 32: 145. 1968. Type: Colombia, Caqueta, Amazon Basin, near Florencia, *FRS66C134* (holotype: UCLA).

DESCRIPTION The sepals are light green to cream-yellow and the petals grade to white over their apical third. The lip interior is green grading to white on the apical third and is striped longitudinally over the apical half with four to seven parallel red-brown lines. The foliage is firm, bright apple-green, and very broad. The lateral sepals reflex backward. The lip is widest near the apex, with low raised veins that radiate from the base to the lip apex. The three-toothed callus is subquadrate with a cavity beneath the apical portion of the callus.

COMMENT Garay (1969) says that *Chondrorhyncha caquetae* is a synonym of *C. rosea* (see for details).

MEASUREMENTS Leaves 18–30 cm long; inflorescence 9–12 cm long; sepals 3.2–4.0 cm long, 1.2 cm wide; petals 3.2 cm long, 1.2–1.5 cm wide; lip 3.2–3.8 cm long, 2.0–2.2 cm wide; column 1.8–2.0 cm long.

ETYMOLOGY Named for the Department of Caqueta in Colombia, where the type specimen was found.

DISTRIBUTION AND HABITAT Known only from Colombia, elevation not recorded.

FLOWERING TIME Not recorded.

Chondrorhyncha hirtzii Dodson, *Icon. Pl. Trop.*, ser. 2, 5: plate 416. 1989. Type: Ecuador, Napo, Baeza, 1500 m, January 1983, *Hirtz 069* (holotype: SEL).

DESCRIPTION The sepals and petals are yellow-white; the lip varies from solid dark red to yellow-white and is flushed and striped with red. The concave dorsal sepal is semierect while the lateral sepals are spreading and recurved. The lip is obovate, widest near the middle or lateral lobes, bilobed at the apex (margin notched), and flared for the apical fourth of the lip. The apical portion of the lip has slightly raised ribs or veins. The callus is five-toothed and lies on the surface of lip. The column is terete and flattened on the underside without wings. PLATE 18.

MEASUREMENTS Leaves 22 cm long, 3 cm wide; inflorescence 12 cm long; dorsal sepal 2.5 cm long, 1 cm wide; lateral sepals 3.2 cm long, 1.1 cm wide; petals 3 cm long, 1.4 cm wide; lip 4 cm long, 2 cm wide; column 2.5 cm long.

ETYMOLOGY Named for Alexander Hirtz, who discovered the species.

DISTRIBUTION AND HABITAT Known from northeastern Ecuador and Colombia, at 1500–1900 m in elevation, in wet montane forest.

FLOWERING TIME January to April.

Chondrorhyncha inedita Dressler &

Dalström, *Orquideología* 23(2): 80. 2004.
Type: Colombia, Ocaña, 1850 m, July 1850,
L. Schlim 34 (holotype: K-L).

DESCRIPTION The original color is problematic
since it is based on a drawing of the type attrib-
uted to Louis Schlim. The sepals are green or
yellow-green, and the petals are white (the specimen label says petals were marked with red).
The lip color was not mentioned in the written description. Another collection on the same
herbarium sheet (*Schlim 1014*) recorded that the
lip was pink. In this drawing of the type specimen used to determine *Chondrorhyncha inedita*,
the lip margins are shown as pink or red-pink,
but the center is shown as very dark, so that
the first impression of flower color is violet. It
may have been colored like that to show the
depth of the throat. The lip is trilobed with the
small midlobe seen as oblong, kidney-shaped,
with warty veins, and the large lateral lobes
are rounded. The callus has five keels basally
which become three keels apically, ending in
three teeth over the lip concavity. The lip concavity begins at mid length proximal to the apical end of the callus and extends both laterally and apically to the beginning of the middle
lobe, so that the apical ends of the callus are
surrounded on three sides by the concavity. The
lip has low warty keels or veins apical to the
concavity.

COMMENT This plant was considered to be a
second collected specimen of *Chondrorhyncha
rosea*, but in comparing the two specimens Robert Dressler and Stig Dalström found many differences. All material of *C. inedita* is from one
dried specimen, and the species has never been
recollected. The concavity at the callus is also
found in *C. caquetae* and *C. rosea*, but apparently
Dressler and Dalström felt the cavity was more
pronounced in *C. inedita*.

MEASUREMENTS Leaves 24–33 cm long, 1.7–2.5
cm wide; inflorescence 6–10 cm long; dorsal
sepal 3 cm long, 0.8–1.0 cm wide; lateral sepals
2.8 cm long, 0.6 cm wide; petals 2.8 cm long,
1.3 cm wide; lip 2.5 cm long, 1.7 cm wide.

ETYMOLOGY Latin *inedita*, unknown or
unpublished.

DISTRIBUTION AND HABITAT Known only from
Colombia, at 1800 m in elevation

FLOWERING TIME July, but not collected since
original specimens were obtained.

Chondrorhyncha manzurii P. Ortíz, *Rev.*

Acad. Colomb. Cienc. 8(90): 57. 2000. Type:
Colombia, Cauca Valley, Calima Lake, Vereda
Varsovia, 1600 m, collected by Julio César
Miranda 1997, cultivated by David Manzur,
October 1999, *P. Ortíz & D. Manzur 1113*
(holotype: COL).

DESCRIPTION The flower is white to green; the
lip is yellow or yellow-green medially with red-
purple spots on the lip skirt and a red suffusion
over the yellow callus. The dorsal sepal and petals parallel the column reflexing at their apexes;
the lateral sepals reflex backward. The oval lip
folds at mid length to form a showy skirt and
the lip margins are crenulate with a slight apical notch. The callus has three keels that begin
at the lip base with the central keel being most
prominent, ending at the plate edge of the callus to form three short teeth. The column is
slightly curved, widest below the slitlike stigma,
with a midline ventral keel present at mid column length. The column apex hoods the clinandrium which is at 90 degrees from the line of
the column length. PLATE 19.

COMMENT This plant is found in an area in
which both *Euryblema andreae* and *Daiotyla maculata* are present and may be a natural hybrid
of the two species. The column of the plant is
most consistent with *Chondrorhyncha* and is the
primary reason I have placed this species in
Chondrorhyncha, but, based on the type drawing,
the viscidium, stipe, and plant foliage are also
consistent with *Chondrorhyncha*. The appearance of the flower also has traits of *Euryblema*.

MEASUREMENTS Leaves 38 cm long, 3.3 cm
wide; inflorescence 1.5 cm wide; dorsal sepal
3.3 cm long, 1 cm wide; lateral sepals 3.7 cm
long, 1.1 cm wide; petals 3 cm long, 1.2 cm

wide; lip 3.6 cm long, 4.2 cm wide; column 2.4 cm long.

ETYMOLOGY Named for David Manzur Macías, of Manizales, Colombia, who cultivated the plant.

DISTRIBUTION AND HABITAT Known only from Colombia, at 1600 m in elevation.

FLOWERING TIME October.

Chondrorhyncha panguensis Dodson ex P. A. Harding, *Orquideología* 25(2). 2008. Type: Ecuador, Morona-Santiago, near Panguí, 1500 m, *Dodson 16003* (holotype: MO!).

DESCRIPTION The sepals and petals are cream-white, the petals have red spots at the apexes, the lip is white with a yellow throat and faint red spots on the lip apex, and the column has red spots. The sepals curl lengthwise their entire length forming tubes; the petals are widest at three-quarter length and recurve at the apex. The lip flares at the crispate apex. The callus has a basal keel and two teeth at its apex, but does not extend to the lateral edges of the lip margins. The column is pubescent on the ventral side. PLATE 20.

COMMENT *Chondrorhyncha panguensis* is similar to *C. velastigui* and is found in the same region in Ecuador. It differs by having an inrolled dorsal sepal, whereas *C. velastigui* has a concave dorsal sepal. The two teeth on the apical callus of *C. panguensis* are not midline, differing from the apical callus of *C. velastigui*, which has a long median tooth at midline with two shorter lateral ones. The foliage of *C. panguensis* is strap-like, similar to *C. hirtzii*, whereas the foliage of *C. velastigui* is more broad and oval, similar to *C. suarezii*. The illustration in *Icones Plantarum Tropicarum*, series 2, plate 413, labeled as *C. andreettae* is *C. panguensis*.

MEASUREMENTS Leaves 20–25 cm long, 1.7–3 cm wide; inflorescence 8–10 cm long, with one or two bracts; dorsal sepal 2.2 cm long, 0.8 cm wide; lateral sepals 3 cm long, 0.8 cm wide; petals 3.1 cm long, 2.2 cm wide; lip 2.5 cm long, 1.5 cm wide; column 1.5 cm long.

ETYMOLOGY Named for the city of Panguí in Ecuador. The Dodson specimen at the Missouri Botanical Garden was labeled *Chondrorhyncha panguensis* but somehow was later considered to be *C. andreettae* by Dodson.

DISTRIBUTION AND HABITAT Known only to eastern Ecuador at 1500–1850 m in wet montane forest.

FLOWERING TIME January to April.

Chondrorhyncha rosea Lindley, *Orch. Linden.*: 13. 1846. Type: Venezuela, Merida, Jajá 1500 m, July 1842, *Linden 651* (holotype: K); type species of *Chondrorhyncha*.

DESCRIPTION The sepals and petals are yellow-green to green, the petals are white, the lip is white with a basal yellow suffusion and with rose-pink or red spots or suffusion over the yellow callus and the apical lip, the color pink becoming sparse at the lip margins. The column is white sometimes with fine red spots. The petals are widest at their apical half (spatulate somewhat). The lip is fleshy, elliptic, and notched at the apex. The callus is three-ridged, raised apically, continuing in three veins to the recurved apex of the lip. PLATE 21.

COMMENT *Chondrorhyncha rosea* has been collected rarely in Venezuela but also has been found in Colombia. A drawing of the type specimen shows the three-toothed callus and the sulcus below it, but does not show the entire lip base so there is no way of knowing whether keels are present or not. Galfrid Dunsterville drew the species with no basal keels, just a raised area at lip mid length. Plate 21 in the present volume shows at least veining from the basal portion of the lip prior to the raised central callus. If this veining is another author's keels, then perhaps *C. rosea*, *C. inedita*, and *C. caquetae* are synonyms. The illustration of *C. rosea* in plate 517 of *Icones Plantarum Tropicarum* (Dodson and Vásquez 1982), showing a plant from Bolivia, has been corrected to *C. fosterae (Ixyophora)*.

MEASUREMENTS Leaves 26 cm long, 1.5–3.5 cm wide; inflorescence 11 cm long; dorsal sepal 2.3–2.5 cm long, 1 cm wide; lateral sepals 3.2–3.5 cm long, 0.8 cm wide; petals 2.6 cm long, 1.2 cm wide; lip 3 cm long, 2.2 cm wide; column 2.2–2.5 cm long.

ETYMOLOGY Latin *roseus*, reddening, referring to the color of the lip or the entire flower.

DISTRIBUTION AND HABITAT Known from Venezuela and Colombia, at 1500–1800 m in elevation, in wet montane cloud forest.

FLOWERING TIME March to July.

Chondrorhyncha suarezii Dodson, *Icon. Pl. Trop.*, ser. 2, 5: plate 418. 1989. Type: Ecuador, Napo, Misahualli, collected by David Neill, 500 m, 9 February 1987, *Dodson 16999C* (holotype: QCNE).

DESCRIPTION The sepals are pale green and the petals and lip are white. The dorsal sepal is erect and concave, the longitudinally inrolled lateral sepals reflex back, and the broad petals reflex only apically. The lip is obovate, obtuse, and flared at the apex. The medial callus is concave, five-lobed with a low central rib and two lateral ribs on each side; the inner rib is equal in length to the central rib and the outermost rib is shorter and retuse at its apex. The column is clavate and sulcate on the underside. The stigma is slitlike. PLATE 22.

COMMENT The leaves of *Chondrorhyncha suarezii* and *C. velastigui* are markedly wider than the typical *Chondrorhyncha* leaves. Both species have an outward appearance similar to *Ixyophora*, but lack the long column foot of *Ixyophora*. Neither of these *Chondrorhyncha* species has been placed using DNA data.

MEASUREMENTS Leaves 18 cm long, 5 cm wide; inflorescence 8 cm long; dorsal sepal 1.5 cm long, 0.8 cm wide; lateral sepals 2 cm long, 0.7 cm wide; petals 2.2 cm long, 1 cm wide; lip 2 cm long, 1.2 cm wide; column 1.1 cm long.

ETYMOLOGY Named to honor Alejandro Suarez, who collected and illustrated the species.

DISTRIBUTION AND HABITAT Known only to Ecuador, at 450–500 m in elevation, in montane wet forest.

FLOWERING TIME Sporadically most of the year

Chondrorhyncha velastiguii Dodson, *Icon. Pl. Trop.*, ser. 2, 5: plate 419. 1989. Type: Ecuador, Tungurahua, Río Negro, 1300 m, 11 March 1963, *Dodson & Thien 2358* (holotype: SEL).

DESCRIPTION The sepals are pale green, and the petals and the lip are white with maroon spots. The concave dorsal sepal is erect, the lateral sepals recurve back and inroll longitudinally. The petals reflex only at the apex. The lip is obovate, obtuse, and flared at the apex. The callus is a transverse plate ending at one-third the lip length, with a central rib and two shorter lateral ribs. The column is clavate and sulcate on the underside. The stigma is slitlike. PLATE 23.

COMMENT The leaves of *Chondrorhyncha suarezii* and *C. velastiguii* are markedly wider than the typical *Chondrorhyncha* leaves. Both species have an outward appearance similar to *Ixyophora*, but lack the long column foot of *Ixyophora*. Neither of these *Chondrorhyncha* species has been placed using DNA data.

MEASUREMENTS Leaves 18 cm long, 5 cm wide; inflorescence 8 cm long; dorsal sepal 2 cm long, 0.8 cm wide; lateral sepals 3 cm long, 0.8 cm wide; petals 3.2 cm long, 1 cm wide; lip 3.5 cm long, 2.2 cm wide; column 2 cm long, 0.5 cm wide.

ETYMOLOGY Named to honor Segundo Velastigui, who helped collect the first plant of this species.

DISTRIBUTION AND HABITAT Known from eastern Ecuador, at 1050–1300 m in elevation, in wet montane cloud forest.

FLOWERING TIME Most of the year.

Chondroscaphe (Dressler) Senghas & Gerlach, *Orchideen* (Schlechter) 1/B(27): 1655. 1993.

Chondroscaphe was established first as a section of *Chondrorhyncha* in 1983 by Robert Dressler, who felt the species included had enough characteristics to remain within *Chondrorhyncha* but were distinct enough to warrant their own section. When *Chondroscaphe* was a section of *Chondrorhyncha*, it included the fringed-lipped *C. flaveola* group and the *C. bicolor* group, both of which shared a second thickening below the basal callus. Senghas and Gerlach (1992) raised *Chondroscaphe* to the genus level but left the *C. bicolor* group in *Chondrorhyncha*, including only the fringed-lipped species within the genus *Chondroscaphe*. Molecular data analyzed by Whitten et al. (2005) confirms that indeed the original section of *Chondroscaphe* is distinct from *Chondrorhyncha*, and the *C. bicolor* group is congeneric with the fringed group, as proposed by Dressler.

Work with this group of species has been difficult as the type description of *Chondroscaphe bicolor* is vague and the type specimen is poorly preserved (Dressler 2001, Pupulin 2005). Also many drawings of the types of the *Chondroscaphe* species have been done by artists who have paid more attention to the frilled lips, leaving off other details that were not written down. This leaves those trying to sort them out, short of doing a taxonomic revision, frustrated. As dried specimens, these species are reputedly difficult to work with. Franco Pupulin and Robert Dressler are working with this group and we await their revision. According to Pupulin (2005), they have many collected plants that don't match the previously described species, but because of the question of what *C. bicolor* may or may not be, these plants remain undescribed at this time.

The leaves of *Chondroscaphe* are long, narrow, and thin in cross section. Inflorescences are one-flowered and emerge from the base of the leaf sheaths. The inflorescence is semierect to laxly pendent, so that the flower is well displayed. The lateral sepals are flat in some species, and tubular and recurved in others; in some species they reflex back toward the ovary, while in other species they do not reflex. The lip disc is thick and rigid. There is a basal or sub-basal two-lobed callus (or a callus of two lamellae or keels), which is generally long, running about half the lip length, with a central sulcus or depression medially between the two lobes. There is another thickening of the lip disc apical to the basal callus, which some authors call a second callus, while others just call it a thickening. The lip is tubular, deflexing to varying degrees at its apex. A few of the nonfringed lipped species have a transverse fold in the lip lateral to the apex of the basal callus that creates a forward-facing surface to the lateral surface of the lip. The column has a pair of teeth or lobules that parallel or clasp the rostellum. The column foot may or may not have a ventral tooth at its base and the column foot is of variable length compared to the column, depending on the species. Only two species have true wings on the column. The pollinarium has a distinct well-developed stipe that is attached to the upper surface of the viscidium rather than the edge. The four pollinia are narrow and unequal, arranged in pairs.

Chondroscaphe species could be divided into the yellow-flowered ones of the valleys of the South American Andes, and the white-flowered ones with maroon spots that have various forms in Central America. There are two species of white *Chondroscaphe* in South America, *C. merana* in Ecuador and *C. plicata* in Peru, and perhaps another species that has been misidentified as *C. bicolor* reported to be from Colombia. The list of species is not long, and one would think this genus, with its showy flowers that are open, should be easy to differentiate.

I have taken many pictures of specimens of this genus and I think I can say that no two are the same. After taking these pictures, I have torn the plants apart and taken pictures of the parts, sometimes with measuring sticks. Even with this amount of information at hand, I find myself pondering over which species is which. In the key and descriptions that follow, I tried to emphasize what I see as important traits for segregating the photographs in my collection. If you only try to match pictures without tearing the flower apart, you may be led astray.

Etymology
Latin *chondroideus*, hard or toughlike cartilage, and *scaphe*, boat-shaped, referring to the shape and texture of the lip.

List of the Species of *Chondroscaphe*

*Chondroscaphe amabilis**

*Chondroscaphe atrilinguis**

Chondroscaphe bicolor

*Chondroscaphe chestertonii**

Chondroscaphe dabeibaensis

*Chondroscaphe eburnea**

Chondroscaphe embreei

*Chondroscaphe endresii**

*Chondroscaphe escobariana**

*Chondroscaphe flaveola** (type species of *Chondroscaphe*)

Chondroscaphe gentryi

*Chondroscaphe laevis**   *Chondroscaphe plicata*

Chondroscaphe merana   *Chondroscaphe yamilethae*

* Molecular sampling confirms placement of this species within the genus
  Chondroscaphe.

Key to the Species of *Chondroscaphe*

1a. Lip fimbriate, flower yellow . . . . go to 2

1b. Lip not fimbriate, flower white, cream-white, or white-green. . . . . . . go to 8

2a. Column straight or mostly straight (excluding the bend at the junction
    of the column and its foot) . . . . go to 3

2b. Column obviously bent, arched, or curved down, or with a dorsal hump
    (in addition to the bend at the column foot) . . . . go to 7

3a. Basal callus composed of two wide, apically rounded lamellae that are
    widest at the lip base and form a shelf extending to the lateral side walls
    of the lip with a central notch or sulcus . . . . go to 4

3b. Basal callus composed of two long lamellae with toothlike apexes . . . .
    . . . . go to 5

4a. Lip more than 2 cm wide . . . . *C. flaveola*

4b. Lip 2 cm wide or less . . . . *C. embreei*

5a. Lip folded lengthwise on itself, apical (second) lip callus with a central
    prominence dissected by a central vein. . . . . go to 6

5b. Lip not folded on itself, apical lip callus either simple or not dissected
    deeply by veins . . . .*C. chestertonii*

6a. Apical callus with two simple lobes . . . .*C. amabilis*

6b. Apical callus with a central keel and three apical teeth . . . . . .*C. dabeibaensis*

7a. Callus apical end raised two knobs . . . . *C. escobariana*

7b. Callus apical end barely raised off the lip. . . . . *C. gentryi*

8a. Petals obovate-spatulate (widest at three-quarters length, narrow at base)
    . . . . go to 9

8b. Petals elliptic (widest at half length or below) . . . . go to 12

9a. Lateral sepals reflexed back, flower white with purple almost black spots
    on lip, Costa Rica. . . . . *C. bicolor*

9b. Lateral sepals not reflexed back, flower white with red spots on lip, yellow
    suffusion over apical callus . . . . go to 10

10a. Basal callus apical end elevated or curved up, Panama . . . . *C. eburnea*

10b. Basal callus apical end not curved up . . . . go to 11

11a. Callus extends apically to more than half lip length, Peru. . . . . . . *C. plicata*

11b. Callus extends to less than half lip length, Ecuador . . . .*C. merana*

12a. Column flat on dorsal surface, lateral sepals not reflexed back, flower
    white, red spots on lip, Costa Rica . . . . *C. yamilethae*

12b. Column rounded on upper surface . . . . go to 13

13a. Lip three-lobed, lateral lobes with forward-facing surface, flower cream-
     white, spots on callus yellow and red, Costa Rica. . . . .*C. endresii*

13b. Lip one-lobed . . . . go to 14

14a. Lip 2.3 cm long and wide, lateral sepals wide, flower green-cream, spots
     on lip dark red, Costa Rica. . . . . *C. laevis*

14b. Lip 3.8 cm long, 2.6 cm wide, lateral sepals narrow, flower white, spots
     on callus dark maroon, Panama and Costa Rica . . . .*C. atrilinguis*

Chondroscaphe amabilis (Schlechter)

Senghas & G. Gerlach, *Orchideen*
(Schlechter) 1/B(27): 1658. 1993. Basionym:
Chondrorhyncha amabilis Schlechter, *Repert.
Spec. Nov. Regni Veg.* 7: 162. 1920. Type:
Colombia, Cauca (holotype: B, destroyed).

DESCRIPTION The sepals, petals, and lip are
yellow, the lip is a deeper yellow medially with
red-brown distinct spots that extend from the
lip base over both the basal callus and the apical callus onto the deflexed lip face. The dorsal
sepal is erect, the lateral sepals reflex slightly
and are longitudinally concave at their base.
The petals are wider than the sepals and have
fimbriate margins. The base of the lip is narrow with upturned lateral edges. The apical lip
has a midline vein apical to the basal callus, the
lateral lip margins are entire (not fimbriate),
the apical margins are markedly fimbriate. The
deeply notched lip deflexes apical to the second
callus (thickening) and often becomes markedly infolded on itself longitudinally so that
the lip apex appears narrow, with the most apical portion of the lip flaring again. The bilobed
basal callus is composed of two long narrow
lamellae. The second or apical callus is three-
veined, the veins extending down to the lip
apex and apical notch as one vein. The column
is suberect, becoming slightly wider toward the
apex. There is no basal ventral tooth mentioned
or shown in the type description. PLATE 24.

COMMENT In Rudolf Schlechter's drawing of
the type, which is now the lectotype, the callus
shows two lamellae that are distinct on their
lateral sides but fused in the midline. The second callus, as it is drawn, is difficult to inter-
pret unless you look at actual pictures of specimens consistent with *Chondroscaphe amabilis*.

This apical, or second, callus is a hard prominence of varying size, depending on the specimen, with three veins; the midvein makes a
fairly significant cut in the prominence so it is
easy to see if you look for it. The drawing shows
a lip that is twice as long as it is wide, though
the measurements Schlechter gives in the text
don't agree with this ratio. Also shown in the
picture is a prominent vein that extends down
the midline of the lip to the notch.

    Chondroscaphe amabilis generally has a large
light yellow flower with a lip that folds on
itself longitudinally. It is popular in cultivation because it is showy. Because it is frequently
brought to shows, I have photographed many
specimens of this species, or at least specimens
that fit the criteria for this species. I think I can
say that I have never seen two that are exactly
the same. Please remember this when you compare your specimen with Plate 24 in this volume. Also remember that the measurements
following are those of the type specimen and
yet size of the floral segments is variable. A
drawing of this species can be found in *Reperto-
rium Specierum Novarum Regni Vegetabilis* (1929:
57, plate 58, drawing 222).

MEASUREMENTS Of the type specimen: Leaves
23–30 cm long, 2.2–3.0 cm wide; inflorescence
12–13 cm long; dorsal sepal 3 cm long; lateral
sepals 4 cm long; petals 3 cm long, 0.9 cm wide;
lip 3.5 cm long, 2.2 cm wide; column 1.2 cm
long.

ETYMOLOGY Latin *amabilis*, loveable. No lie!

DISTRIBUTION AND HABITAT Known from
Colombia and Ecuador.

FLOWERING TIME Not recorded.

Chondroscaphe atrilinguis Dressler,

Orquideología 22(1): 16. 2001. Type: Panama, Bocas del Toro, flowered in cultivation 8 September 2000, *R. L. Dressler 6289* (holotype: MO!).

DESCRIPTION The sepals are pale green, the petals and lip are cream with dark red almost black spots on the callus and throat of the lip. The lip apex is pointed, sometimes sharply, with undulate, crenulate, or subfimbriated margins. The lateral lobes of the lip encircle the column and have a prominent transverse keel lateral to the apex of the basal callus causing a fold in the lip. The basal callus has a glandular surface and free rounded apexes. The apical callus is rounded, oblong, and rugulose. The stout column, with a rounded dorsal surface, has slightly spreading wings below the stigma, and its ventral surface is puberulent. PLATE 25.

COMMENT The photograph labeled *Chondrorhyncha bicolor* in Senghas and Gerlach (1992, figure 1514) shows *C. atrilinguis*. The drawing labeled *C. bicolor* in Jenny (1983) is *C. laevis*.

MEASUREMENTS Leaves 18–35 cm long, 2.2–2.4 cm wide; inflorescence 2.5–4.5 cm long; dorsal sepal 3.0–3.3 cm long, 1.0–1.1 cm wide; lateral sepals 3.3–4.2 cm long, 0.75–1.10 cm wide; petals 2.9–3.1 cm long, 1.3–1.4 cm wide; lip 3.8–4.5 cm long, 2.6–3.1 cm wide; column 1.8–2.0 cm long, 0.6 cm wide.

ETYMOLOGY Latin *atri*, black or dark, and *lingulatus*, tongue-shaped, after *boca negra*, black mouth, the name for the species in Panama, referring to the blackish color around the tonguelike callus.

DISTRIBUTION AND HABITAT Known from Panama and Costa Rica, at 800 m in elevation.

FLOWERING TIME Sporadically from May to September.

Chondroscaphe bicolor (Rolfe) Dressler,

Orquideología 22(1): 22. 2001. Basionym: *Chondroscaphe bicolor* Rolfe, *Bull. Misc. Inform. Kew*: 393. 1894. Type: Costa Rica, *Pfau 717* (holotype: K).

DESCRIPTION The sepals and petals are white, the lip is white with purple markings on the middle and base of the lip. The sepals and petals are erect, oblong, and pointed. The lip is obscurely trilobed, with the kidney-shaped middle lobe having crenulate margins. The original description says the lip has no callus. The column is clavate.

COMMENT Due to the poor condition of the type specimen, this species has caused much debate as to whether it still exists or whether other species more recently described should be included as synonyms of *Chondroscaphe bicolor*. What authors have done, and perhaps correctly, is moved on and described new species with good descriptions and specimens, leaving *C. bicolor* a mystery, without living material. Pupulin (2005) writes that the basal callus of the type specimen is bilobed with two triangular rounded apical teeth, and from the watercolor of the type (*Pfau 717*, K), one sees divergent pararostellar lobes on the column, a feature which is said to be diagnostic. The photograph labeled *C. bicolor* in Senghas and Gerlach (1992, figure 1514) shows *C. atrilinguis*. The drawing labeled *C. bicolor* in Jenny (1983) is *C. laevis*.

MEASUREMENTS Leaves 40–65 cm long, 2.0–2.7 cm wide; inflorescence 5.4 cm long; dorsal sepal 2.7 cm long, 1 cm wide; lateral sepals 2.7 cm long, 0.6 cm wide; petals 2 cm long, 0.7 cm wide; lip 2.4 cm long, 1.8 cm wide; column 1.4 cm long.

ETYMOLOGY Latin *bi*, two, and *colur*, color, referring to the flowers having two colors.

DISTRIBUTION AND HABITAT Known only for certain from Costa Rica. Other reports of ranges for the species being from Costa Rica to Ecuador are most likely of other species.

FLOWERING TIME Not recorded.

Chondroscaphe chestertonii (Reichenbach f.) Senghas & Gerlach, *Orchideen* (Schlechter) 1/B(27): 1657. 1993. Basionym: *Chondrorhyncha chestertonii* Reichenbach f., *Gard. Chron.*: 648. 1879. Type: Colombia (holotype: W).

SYNONYM
Chondrorhyncha chestertonii var. *major* hort. ex Cogniaux, *Dict. Icon. Orch.*: plate 1. 1901.

DESCRIPTION The flowers are yellow with dark red spots at the base of the lip. The petal margins are slightly crispate. The undulated lip is deeply lacerate with an apical notch. The apical margin of the lip callus has two points or teeth, often with a third tooth that is below and between the apical ends of the lamellae of the callus. The second callus, or thickening, is simple without knobs, lines, or deep veins. The column is straight, widening apically, with a ventral tooth at the column base. PLATE 26.

COMMENT Specimens I have seen tend to have sepals and petals that are less full with more space between them than those of *Chondroscaphe amabilis* or other species. The flowers are smaller than the other showier species also. *Chondroscaphe chestertonii* is bigger and more fimbriate than *C. flaveola*. Francisco Villegas bought a plant of *C. chestertonii* at the Pereira, Colombia, exhibition in 2006; this plant has thick teeth extending laterally from the basal half of the basal callus, but all other traits are similar to those of *C. chestertonii*. The illustration in *Icones Plantarum Tropicarum* (1980: series 1, plate 23) labeled originally as *C. chestertonii* shows *C. dabeibaensis*.

MEASUREMENTS Leaves 20–25 cm long, 1.0–1.5 cm wide; inflorescence 12–16 cm long; dorsal sepal 3 cm long, 0.8 cm wide; lateral sepals 4.2 cm long, 1.2 cm wide; petals 4 cm long, 2 cm wide; lip 4 cm long, 3.5 cm wide; column 1.9 cm long.

ETYMOLOGY Named for Chesterton.

DISTRIBUTION AND HABITAT Known from the Cauca region of Colombia and from northern Ecuador.

FLOWERING TIME Not recorded.

Chondroscaphe dabeibaensis P. A. Harding, *Orquideología* 25(2). 2008. Type: Colombia, Antioquia, Dabeiba, Chimado, 1600 m, collected by Hector Angarita, July 1992, *C. H. Dodson & R. Escobar 19006* (holotype: MO!).

DESCRIPTION Flower yellow to white-yellow with red-brown spotting on the lip disc and calluses. The sepals are elliptic and the petals are spatulate with crenulate to slightly fimbriate margins. The lip deflexes at mid length with fimbriate margins and with a midline longitudinal fold and an apical notch. The basal callus is bilobed, fused somewhat centrally with a central keel, all ending apically in three acute teeth. The apical callus is short, consisting of a prominent knob distal to the basal callus teeth, which forms a transverse ridge. The column is straight, slightly upturned at the apex, with obvious acute wings lateral to the stigma, and has a ventral tooth at the base of the column foot. PLATE 27.

COMMENT Plate 21 in *Icones Plantarum Tropicarum*, Ecuador, illustrates a specimen, *Dodson & Thien 1595*, that has changed names over the years. It was originally used to illustrate *Chondrorhyncha chestertonii*. When Calaway Dodson described *C. embreei* he listed the specimen in plate 21 as a paratype. In *Native Orchids of Ecuador* (Dodson and R. Escobar 1993: 119), the same drawing (plate 21) shows up as *C. amabilis*, with a picture of *C. amabilis* below it. This plate 21 depicts many of the characteristics Rudolf Schlechter has in his *C. amabilis*, but the specimen in the plate lacks the deep notch at the apex of the lip, the lip length-to-width ratio is different, being more ovate than oblong, the column has acute wings, and the fused lamellae of the basal callus have a midline keel that ends in an acute small tooth or point.

I photographed two different plants of *Chondroscaphe* in Gustavo Aguirre's collection, Colombia, with these traits; one of them is reproduced as Plate 27 in the present volume. I probably wouldn't have thought twice about these specimens, except that Rudolf Jenny of Switzerland sent a photograph of a specimen with the same characteristics. Further close looking revealed another specimen in *Native Colombia Orchids* (R. Escobar 1994), where the photograph on page 79 is labeled *C. chestertonii*, but the one on page 80 is labeled *C. amabilis*. Clearly the two photographs were reversely labeled. The photograph on page 79 shows the folded lip, the prominence that is deeply cut by the central vein, and the basal callus with the midline keel and third tooth.

A visit to the herbarium at the Missouri Botanical Gardens revealed two specimens labeled *Chondrorhyncha dabeibaensis*, determined by Dodson, which makes me think Dodson thought he had a new species at one point but for whatever reason never wrote it up.

MEASUREMENTS Leaves 25 cm long, 1.7 cm wide; inflorescence 10 cm long; dorsal sepal 2.2–3.0 cm long, 0.5 cm wide; lateral sepals 4 cm long, 0.5 cm wide; petals 3.2–3.5 cm long, 0.8–1.2 cm wide; lip 2.5–4.0 cm long, 2.5–3.2 cm wide; column 2.2 cm long.

ETYMOLOGY Named for the municipality where it was discovered.

DISTRIBUTION AND HABITAT Known from Colombia and Ecuador, at 1400–1600 m.

FLOWERING TIME Sporadically throughout the year.

Chondroscaphe eburnea (Dressler) Dressler, *Orquideología* 22(1): 22. 2001. Basionym: *Chondrorhyncha eburnea* Dressler, *Orchidee (Hamburg)* 34(6): 224. 1983. Type: Panama, Coclé, Cerro Caracoral, hills north of Antón Valley, 850 m, 9 July 1982, *R. L. Dressler 6070* (holotype: US; isotype: PMA).

DESCRIPTION The sepals are green-cream, the petals, lip, and column are cream-yellow, the basal callus and the base of the lip's midlobe second callus are yellow with red-purple spots, and the throat of the lip has red-purple spots. The dorsal sepal is concave and hooked, and the lateral sepals are markedly concave with pointed apexes. The petals are wider than the sepals and are pointed. The base of the lip is

cuneate, the lip is weakly three-lobed, and the apical margin is crisped. The basal callus is 6 mm wide ending in two rounded teeth that are elevated at the lip base apically, so that the whole callus curves upward apically. The column is widest above the stigma, with a tooth (Dressler says umbo) on the column foot. PLATE 28.

COMMENT There is a photograph of the species with the original type description.

MEASUREMENTS Leaves 15–37 cm long, 2.5–4.0 cm wide; inflorescence 5 cm long; dorsal sepal 2.5–2.7 cm long, 0.9–1.0 cm wide; lateral sepals 3.5–3.8 cm long, 0.9–1.0 cm wide; petals 2.7–3.0 cm long, 1.0–1.2 cm wide; lip 3.0–3.5 cm long, 2.2–2.7 cm wide; column 1.4–1.5 cm long, 0.7 cm wide.

ETYMOLOGY Latin *eburneus*, the color of ivory (white with a yellow tinge), referring to the flower color.

DISTRIBUTION AND HABITAT Known only from Panama, at 850 m.

FLOWERING TIME July.

Chondroscaphe embreei (Dodson & Neudecker) C. Rungius, *Orchidee*, Suppl. 3: 16. 1996. Basionym: *Chondroscaphe embreei* Dodson & Neudecker, *Orquideología* 19(1): 82. 1993. Type: Ecuador, Pichincha, San Miguel de los Bancos, 1500 m, 25 September 1980, *Andreetta 1209* (holotype: SEL).

DESCRIPTION The sepals and petals are yellow-green to white-yellow, and the lip is white to yellow-orange with numerous small spots of red-brown in the throat and on the basal callus. The petals are crenulate and irregular but not fimbriate. The lip is bilobed at the apex (notched at the apex), deeply and irregularly fimbriate on the apical lip margins, sparing the apical notch and the basal portion of the lip. The lip is tubular and parallel to the column for the basal third, flaring and recurved abruptly for its apical half. The basal two-toothed rectangular callus is raised, widest at the base, and extends one-third to one-half the length of the lip, and the forward callus is transverse with a hard central triangular apical projection. The clavate column is straight and slightly upturned at the apex. The column lacks an upturned small tooth at the apex of the column foot ventral surface. PLATE 29.

COMMENT The species was initially identified as *Chondroscaphe flaveola* and is treated as such in plate 23 of *Icones Plantarum Tropicarum* (1980, ser. 1). The plate published with the original description of *C. embreei* is said in the article to be plate 23 in *Icones Plantarum Tropicarum*, labeled *C. flaveola*, but the wrong plate was inserted with the description—plate 21, which in the article is listed as a paratype of *C. embreei*. Stig Dalström was kind enough to look closely at the type specimen of *C. embreei* and confirm that it looks like the illustration labeled *C. flaveola*, plate 23 of the *Icones Plantarum Tropicarum*, and not plate 21.

Chondroscaphe flaveola and *C. embreei* are very similar, but differ in that *C. flaveola* is much bigger and *C. embreei* has the V-shaped prominence that sticks out like a nose or knob from the lip surface apical to the basal callus. I do not believe the triangular projection is diagnostic of this species, as *C. amabilis* and *C. dabeibaensis* have this also, but *C. amabilis* and *C. dabeibaensis* have veins that deeply dissect the prominence and have lips which fold in half lengthwise. The type drawing of *C. flaveola* by Louis Schlim seems to have a prominence in the apical region, though of the plants of *C. flaveola* that I have seen, if they have this prominence, it doesn't project out, but is rather a raised portion only.

Plate 29 in the present volume is one I photographed in Colombia, where I took the flower apart and photographed it and its pieces, so I am fairly sure this is a good match for plate 23 of *Icones Plantarum Tropicarum*, except that this specimen has a tooth on the ventral column foot. The description of the type of *Chondroscaphe embreei* says this tooth is lacking in the type specimen.

MEASUREMENTS Leaves 32 cm long, 2.5 cm wide; inflorescence 12 cm long; dorsal sepal 2.8 cm long, 1 cm wide; lateral sepals 4 cm long, 1 cm wide; petals 3 cm long, 1.1 cm wide; lip 3.5 cm long, 2.2 cm wide; column 2.5 cm long.

ETYMOLOGY Dedicated to Alvin Embree, who participated in the collection of the type specimen.

DISTRIBUTION AND HABITAT Known from Colombia and Ecuador, between 950 and 2000 m in elevation.

FLOWERING TIME September.

Chondroscaphe endresii (Schlechter)

Dressler, *Lankesteriana* 3: 28. 2002. Basionym: *Chondroscaphe endresii* Schlechter, *Repert. Spec. Nov. Regni Veg.* 17: 14. 1921. Type: Costa Rica, La Palma, 1922, *C. Wercklé 119* (holotype: W).

DESCRIPTION No color is given in the original description though the drawing shows spots (could be warts) on the callus and the lip inner surface, and a large dark spot or spots on the basal portion of the lip surface. The sepals and petals are conduplicate with a central vein. The lateral sepals do not reflex back severely. The petals have undulate margins. The lip is obscurely three-lobed, the lateral lobes are minimal, the midlobe is notched at the apex, erose or denticulate, but not fimbriate, and recurved for the apical third. The basal callus is narrow basally, widening at mid callus length, narrowing again and becoming bilobed with rounded apexes. The second thickening is apical to the basal callus at the point of deflection. The column is widest and has obtuse prominences but not wings lateral to the stigma, and the dorsal surface of the column is rounded. PLATE 30.

COMMENT *Chondroscaphe endresii* has been determined to be the identity of most of the species previously identified as *C. bicolor*, and *C. bicolor* is still unmatched to plants seen in Mesoamerica. There is a lovely photo of this species by Franco Pupulin at www.jardinbo tanicolankester.org. It shows a cream-white flower, a cream-white lip with red spots on the callus, and a yellow apical thickening with red spots.

MEASUREMENTS Taken from a photograph of the holotype drawing: Leaves 24 cm long, 2 cm wide; inflorescence 3–4 cm long; flower 3 cm long, 3 cm wide; sepals and petals under 2 cm long; lip 1 cm long, 1 cm wide.

ETYMOLOGY Named for A. R. Endrés.

DISTRIBUTION AND HABITAT Known from Costa Rica and Panama, at 950–1200 m in elevation.

FLOWERING TIME October to November.

Chondroscaphe escobariana (Dodson & Neudecker) C. Rungius, *Orchidee*, Suppl. 3: 16. 1996. Basionym: *Chondrorhyncha escobariana* Dodson & Neudecker, *Orquideología* 19(1): 47. 1993. Type: Colombia, Antioquia, Urrao, El Llavero, collected in December 1989 by Manual Zapata, 1500 m, flowered in cultivation at El Hatillo near Medellín, 9 August 1992, *R. Escobar 4032* (holotype: JAUM; isotype: RPSC).

DESCRIPTION The sepals and petals are yellow-green, the lip is yellow-orange with numerous small spots of red-brown in the throat and on the basal callus. The petals are irregularly lacerate on their margins. The lip is bilobed at the apex, deeply and irregularly lacerate or fimbriate on the apical margins, tubular and parallel to the column for the basal quarter of the lip length, recurving abruptly for the apical half of its length. The bilobed callus is shallow at its base and markedly raised at its apex so that from the front there appear to be prominent knobs in the convex portion of the lip. The forward callus on the blade of the lip has four parallel, shallow, subwarty ribs. The column is short, geniculate, blunt at the apex, without wings, and the basal half is swollen on the ventral side. The column foot has a tooth. PLATE 31.

MEASUREMENTS Leaves 40 cm long, 2.7 cm

wide; inflorescence 15 cm long; dorsal sepal 3.7 cm long, 1 cm wide; lateral sepals 4.5 cm long, 1 cm wide; petals 4.1 cm long, 2 cm wide; lip 4.2 cm long, 4.4 cm wide; column 1.8 cm long.

ETYMOLOGY Named for Rodrigo Escobar R., who photographed and studied the orchids of Colombia, and who drew Calaway Dodson and Tilman Neudecker's attention to the species.

DISTRIBUTION AND HABITAT Known only from Colombia, at 1500 m in elevation.

FLOWERING TIME December, recorded as August in cultivation.

Chondroscaphe flaveola (Linden & Reichenbach f.) Senghas & Gerlach, *Orchideen* (Schlechter) 1/B(27): 1656. 1993. Basionym: *Zygopetalum flaveolum* Linden & Reichenbach f. ex Reichenbach f., *Ann. Bot. Syst.* 6: 662. 1863. Type: Colombia, Ocaña, *Schlim s.n.* (holotype: W); type species of *Chondroscaphe*.

SYNONYMS
Stenia fimbriata Linden & Reichenbach f., *Gard. Chron.*: 1313. 1868.
Chondrorhyncha fimbriata (Linden & Reichenbach f.) Reichenbach f., *Refug. Bot.* 2: plate 107. 1872.
Kefersteinia flaveola (Linden & Reichenbach f.) Schlechter, *Repert. Spec. Nov. Regni Veg. Beih.* 7: 266. 1920.
Chondrorhyncha flaveola (Linden & Reichenbach f.) Garay,, *Bot. Mus. Leafl.* 21: 256. 1967.
Chondroscaphe fimbriata (Linden & Reichenbach f.) Dressler, *Orquideología* 22(1): 22. 2001.

DESCRIPTION The flower is transparent light to deep yellow and the callus is a deeper yellow with red warts and spots. The lateral sepals reflex back and are longitudinally slightly concave. The petals are fimbriate but not to the degree of the lip. The lip is fimbriate all around its margins except the most basal portion. The bilobed callus is composed of two broad lamella that extend onto the lateral side lobes of the lip and form a shelf apically with a warty apical surface. The secondary thickening or callus is veined and sometimes toothed in the center. The column is fairly straight except for the column foot and remains clavate at the apex rather than sweeping upward. PLATE 32.

COMMENT Louis Schlim's drawing of the type specimen has petals which are crenulate and not fimbriate. The original (in French) description of *Stenia fimbriata* says the flower is uniform yellow to sulfur and that the basal lip center has a circular series of small crimson-blood spots. The petals are crenulate (*crénelés* is the word used, not *frange*, which is used to describe the lip) at their edges. The lip is large with a fine fringe all around the median lobe. This begs the debate whether *Chondroscaphe fimbriata* and *C. flaveola* are synonyms, but as both *Stenia fimbriata* and *Zygopetalum flaveola* seem to cite the same specimen as the type, they should be considered the same species. The next question would be, Are all the *C. flaveola* with fimbriate petals a different species? I think this trait must just be variable and would defer to the similarity of the callus of these forms as being a unifying trait.

Günther Gerlach (pers. comm.) reports that the illustration in Schlechter's (1993: 1656) *Orchideen* labeled *Chondroscaphe flaveola* is incorrect; the drawing is of a plant from Ecuador. He also says that photographs on the next page are incorrect except for the one in the lower right corner. The incorrect illustration is of a flower with a column that is straight and slightly upcurved at the apex, and with a toothed column foot. The basal callus is elliptic with an notch producing two sharp points apically without more ornamentation apical to the basal callus. I am not sure what species is illustrated but suspect it is *C. chestertonii*.

Plate 107 in Saunders (1869) labeled as *Chondrorhyncha fimbriata* shows a flower from Peru with minimally crenulate petal margins, a notched lip apex, minimal fimbriae on the lip, and a straight clublike column with no tooth at the apex of the column foot. As *C. flaveola* is not noted to be from Peru by other authors, this illustration is also suspect and likely to be *C. amabilis*. Although that species is also

not reported as being from Peru, borders have changed in 150 years and perhaps it was an original citation from Ecuador.

Please note that plate 23 of *Chondrorhyncha flaveola* in *Icones Plantarum Tropicarum* (1980, series 1) has been corrected to *C. embreei*, and plate 27 in *Icones Plantarum Tropicarum* (1989, series 2) is probably *C. plicata*.

Plate 32 in the present volume is provided by Günther Gerlach, who reports it was taken where the original specimen was collected, so it is probably correct. The warts on the callus edges are obvious in the photograph, a characteristic the other species in the genus lack.

MEASUREMENTS Leaves 41 cm long, 2.8 cm wide; inflorescence 11 cm long; dorsal sepal 2.9 cm long, 1.1 cm wide; lateral sepals 4 cm long, 1.2 cm wide; petals 3.2 cm long, 1.5 cm wide; lip 3.8 cm long, 2.9 cm wide; column 1.2 cm long, no column foot.

ETYMOLOGY Latin *flaveolus*, yellowish or pale yellow, referring to the flower color.

DISTRIBUTION AND HABITAT Known from Colombia and Venezuela, flowering in June to October.

FLOWERING TIME Not recorded.

Chondroscaphe gentryi (Dodson & Neudecker) C. Rungius, *Orchidee*, Suppl. 3: 16. 1996. Basionym: *Chondrorhyncha gentryi* Dodson & Neudecker, *Orquideología* 19(1): 49. 1993. Type: Ecuador, Esmeraldas, road from Lita to Cristal, 1450 m, *Dobson & Gentry 17674* (holotype: RPSC; isotype: K).

DESCRIPTION The sepals and petals are cream-yellow to yellow, the lip is cream-yellow with an orange throat and numerous small spots of red-brown in the throat and calluses. The lateral sepals are recurved and spreading. The petals are irregularly lacerate and spreading. The elliptical lip has an apical notch. The lip is deeply and irregularly lacerate on the apical margins, sparing the notch. The lip is tubular and parallel to the column for the basal quarter, flaring and recurved abruptly at mid length. The basal callus is shallow and bilobed at the apex, remaining appressed to the lip, and the forward callus, or thickening, on the blade has several shallow parallel warty ribs. The column is short, geniculate, swollen on the underside of the basal half like *Chondroscaphe escobariana*, blunt at the apex, and lacking a tooth on the column foot. PLATE 33.

COMMENT *Chondroscaphe gentryi* and *C. escobariana* are the only two species in the genus with a markedly bent column, causing the clinandrium to face forward. They differ in that the callus of *C. gentry* lies flat apically whereas the callus of *C. escobariana* curves to almost 90 degrees off the lip floor.

MEASUREMENTS Leaves 38 cm long, 4.7 cm wide; inflorescence 14 cm long; dorsal sepal 4 cm long, 1.1 cm wide; lateral sepals 4.4 cm long, 1.2 cm wide; petals 3.7 cm long, 1.4 cm wide; lip 4.1 cm long, 4.4 cm wide; column 1.8 cm long.

ETYMOLOGY Dedicated to Alwyn Gentry, who participated in the collection of the plant.

DISTRIBUTION AND HABITAT Known only from Ecuador, at 600–1450 m in elevation.

FLOWERING TIME Not recorded.

Chondroscaphe laevis Dressler, *Orquideología* 22(1): 20. 2001. Type: Costa Rica, Alajuela, San Ramón, Alberto Manuel Brenes Biological Reserve, flowered in cultivation at Lankester Botanical Garden, 15 May 1998, *G. Hoffmann s.n.* (holotype: MO; isotype: US).

DESCRIPTION The sepals and petals are greenish cream-yellow, the lip is cream-yellow, the callus and the bases of the lip and column have red-brown spots or stains. The flower is cupped. The relatively wide lateral sepals are held with minimal inrolling at the edges, not forming a long tube. The lip lateral margins are erect, the apical margins are undulate and subcrenulate. The basal callus is covered with branlike scales in its middle, the basal callus is 1 cm long with triangular free apexes. The apical callus, or

thickening, is hemicircular with a few slight creases basally; the remainder of the second callus is smooth. The column, with a rounded dorsal surface, has prominent wings at the stigma, and there is no mention of a tooth at the base of the column foot. PLATE 34.

COMMENT There is a lovely photo of this species by Franco Pupulin at www.jardinbotanicolankester.org. The drawing and picture in Jenny (1983) labeled *Chondroscaphe bicolor* is *C. laevis*. Pridgeon's (1992) *Illustrated Encyclopedia of Orchids* has a photograph labeled *C. bicolor* which I believe is *C. laevis*, but the dorsum of the column looks flat.

MEASUREMENTS Leaves 30 cm long, 1.3–2.0 cm wide; inflorescence 2.0–6.5 cm long; dorsal sepal 1.9 cm long, 0.6–0.7 cm wide; lateral sepals 1.8–2.4 cm long, 0.6–0.7 cm wide; petals 1.5–2.0 cm long, 0.9–1.0 cm wide; lip 2.0–2.3 cm long, 1.8–2.3 cm wide; column 0.9 cm long, 4.8–5.0 cm wide.

ETYMOLOGY Latin *laevis*, smooth, referring to the apical callus.

DISTRIBUTION AND HABITAT Known only from Costa Rica, at 950–1200 m in elevation.

FLOWERING TIME May.

Chondroscaphe merana (Dodson & Neudecker) Dressler, *Orquideología* 22(1): 22. 2001. Basionym: *Chondroscaphe merana* Dodson & Neudecker, *Orquideología* 19(1): 83. 1993. Type: Ecuador, Baños to Puyo, Mera, 1500 m, 17 March 1976, *Luer et al. 891* (holotype: SEL).

DESCRIPTION The flowers are white and the lip has a chocolate-colored blotch in the throat and a few red spots near the apex of the callus. The dorsal sepal and petals form a hood over the column and reflex back at their apexes. The margins of the lateral sepals curve inward and lie parallel to the lip (do not reflex back). The lip is notched at the apex, deeply and irregularly undulate on the apical margins, tubular and parallel to the column for the basal half, flaring and recurved abruptly for the apical third. The callus is rectangular, slightly raised, two-toothed at the apex. The four pollinia are superposed at the apex of an obvious stipe attached to a heart-shaped viscidium. PLATE 35.

COMMENT This species was earlier thought to be *Chondroscaphe bicolor* and was treated as such in plate 414 of *Icones Plantarum Tropicarum* (1989, series 2). *Chondroscaphe merana* and *C. plicata* are very similar, differing in the length the callus extends apically on the lip; the callus of *C. plicata* from Peru covers a larger portion of the lip apically than the callus of *C. merana* from Ecuador. *Chondroscaphe merana* has more red and yellow coloration in the lip throat compared to the dark maroon of *C. plicata*.

MEASUREMENTS Leaves 36 cm long, 2 cm wide; inflorescence 8 cm long, with one or two bracts; dorsal sepal 2 cm long, 0.7 cm wide; lateral sepals 4 cm long, 1 cm long; petals 2.2 cm long, 0.7 cm wide; lip 2 cm long, 2.5 cm wide; column 1.5 cm long.

ETYMOLOGY Named for the location where it was discovered.

DISTRIBUTION AND HABITAT Known only from Ecuador, at 1500 m in elevation, in extremely wet montane cloud forest.

FLOWERING TIME February to April.

Chondroscaphe plicata (D. E. Bennett & Christenson) Dressler, *Orquideología* 22(1): 22. 2001. Basionym: *Chondroscaphe plicata* D. E. Bennett & Christenson, *Brittonia* 46(1): 24. 1994. Type: Peru, Junín, Kivinaki, 1700 m, 26 March 1992, *O. del Castillo ex Bennett 5507* (holotype: MOL).

DESCRIPTION The sepals and petals are pale green-white, the lip is cream-white with dark maroon and yellow markings, the column and anther are cream-white, and the pollinia are yellow. The tubular lip is unlobed, the basal third is constricted and tubular. The lip basal callus consists of two elongate low broad ribs

that converge basally, becoming thicker and free at the oblique tips. The apical callus is verrucose across the middle. The lip apex is recurved, and the apical margins have many pleats. The column is arcuate only slightly, clavate, and the ventral surface is covered with long verrucose processes.

COMMENT Plate 27 in *Icones Plantarum Tropicarum*, Peru (1989, series 2), labeled *Chondrorhyncha fimbriata*, is most likely *C. plicata*, though the measurements of the segments of the flower are longer than those following. *Chondroscaphe merana* from Ecuador and *C. plicata* from Peru are very similar, differing in the length the callus extends apically on the lip; the callus of *C. plicata* covers a larger portion of the lip apically than does the callus of *C. merana*. Furthermore, *C. merana* has more red and yellow coloration in the lip throat compared to the dark maroon almost black of *C. plicata*.

MEASUREMENTS Leaves 15–21 cm long, 1.5–1.9 cm wide; inflorescence 7 cm long; dorsal sepal 1.8 cm long, 0.8 cm wide; lateral sepals 2.6 cm long, 0.7 cm wide; petals 2.3 cm long, 0.8 cm wide; lip 2.4 cm long, 1.8 cm wide; column 1.8 cm long, 0.6 cm wide.

ETYMOLOGY Latin *plicatus*, folded into pleats, referring to the lip and petal apexes.

DISTRIBUTION AND HABITAT Known only to Peru, at 1700 m in elevation, in wet montane forest.

FLOWERING TIME October and November.

Chondroscaphe yamilethae Pupulin, *Vanishing Beauty: Native Costa Rican Orchids* 1: 111. 2005. Type: Costa Rica, Puntarenas, Buenas Aires, Holán, 1200–1300 m, collected by C. Arguedas in 2000, flowered in cultivation 20 April 2003, *F. Pupulin 4701* (holotype: USJ).

DESCRIPTION The sepals and petals are translucent white and the lip is white with red spots on the lateral walls and second thickening. The dorsal sepal reflexes upward at its apex. The petals are parallel to the column and cover it from above. The lateral sepals are concave and held to the sides of the lip in the same direction as the lip, not reflexing back. The lip is obscurely three-lobed, the short midlobe has a notched apex and the midlobe margins are crenulate and shallowly fimbriate. The basal callus is bilobed with pointed apical teeth, and the slightly warty apical callus (second thickening) is rounded and notched. The column is straight without a tooth on the column foot.

COMMENT There is a lovely photo of this species by Franco Pupulin at www.jardinbotanicolankester.org.

MEASUREMENTS Leaves 10–30 cm long, 1.8–4.0 cm wide; inflorescence 4 cm long; dorsal sepal 2.5 cm long, 1 cm wide; lateral sepals 3.8 cm long, 0.9 cm wide; petals 3 cm long, 1.2 cm wide; lip 3.3 cm long, 2.5 cm wide; column 1.9 cm long.

ETYMOLOGY Named for Yamileth González García, president of the University of Costa Rica, in acknowledgment of the strong support given under her direction to the research activities of the Lankester Botanical Garden in Turialba, Costa Rica.

DISTRIBUTION AND HABITAT Known only from Costa Rica, at 1200–1300 m in elevation, in lower montane rainforest on the Pacific slopes of the Cordillera de Talamanca.

FLOWERING TIME April, but probably sporadically throughout the year.

Cochleanthes Rafinesque, *Fl. Tellur.* 1: 45. 1836.

Cochleanthes was first established by Constantine Rafinesque in 1836, using the name *C. fragrans* which was based on the earlier *Zygopetalum cochleare* (which is the type of *Cochleanthes*), and they are both synonyms of the earlier described *Epidendrum flabelliforme*, which has species name priority. This makes the correct name *C. flabelliformis*.

Over time, many species were added and taken away from the genus, depending on the criteria for inclusion in the genus at the time. Schultes and Garay (1959) moved the species of *Warczewiczella* into *Cochleanthes*. Fowlie (1969b) felt there was significant difference between the two and separated them. The same year, Garay (1969) argued that the species should still be in one genus. The species have been in commercial trade as *Cochleanthes* in recent years.

Molecular data reported by Whitten et al. (2005) showed the genus *Cochleanthes* and *Warczewiczella* are indeed separate. Furthermore, the two are separated by other genera that are closer on a molecular level to each genus. *Cochleanthes* is closest to *Stenotyla*. *Warczewiczella* is closer to *Aetheorhyncha*, *Chaubardiella*, *Chondroscaphe*, *Ixyophora*, and *Pescatorea*. The change (*update* is probably the better word) is fairly easy to remember. *Cochleanthes* now includes only two species, the rest that were considered by the general public to be *Cochleanthes* are now in *Warczewiczella*.

The plants are large, almost the size of the plants of *Pescatorea*, and lack pseudobulbs. The foliage lacks the sheaths of *Chondrorhyncha*, the leaves alternating to produce a fanlike appearance. A short rhizome causes the plants to grow in tufts. The leaves are narrow, conduplicate, and possess a midline vein which is dorsally keeled. There is one flower per inflorescence, emerging from the axis of the alternating foliage. The lateral sepals are reflexed (though not as much as *Warczewiczella*). The lateral lobes of the lip are erect but do not enfold the base of the column, creating more of a shelflike extension of the callus than forming a tube as in *Warczewiczella*. The claw of the lip is minimal and thick. The callus is attached solidly to the lip base and lateral portion of the lip. Instead of the radiating network of digitate promontories of *Warczewiczella*, the callus consists of a series of adjacent plates, the apical terminations of which attach to and extend onto the lip, forming the venation of the lip, similar to the callus of *Pescatorea*. The column, which is slightly bent or curved at the apex, has a central ventral longitudinal keel and a distinct tooth at the base of the short column foot. There are four pollinia which are more or less equal. The stipe is minimal and the viscidium is quadrate; the pollinia attach to one side of the viscidium.

    Cochleanthes species prefer shady humid habitats from near sea level up to 1800 meters in elevation. The species range from Costa Rica and Panama down to Peru and Brazil, and the islands of the Caribbean.

Etymology

Greek *kochlias*, spiral-shaped shell, and *anthos*, flower, referring to the shape of a pronounced callus at the base of the lip.

List of the Species of *Cochleanthes*

Cochleanthes aromatica*
Cochleanthes flabelliformis* (type species of *Cochleanthes*)
* Molecular sampling confirms placement of this species within the genus *Cochleanthes*.

Key to the Species of *Cochleanthes*

1a. Lip widest apically, sepals and petals white, lip white with purple medially, margins white . . . . *C. aromatica*
1b. Lip square or widest basally, sepals and petals green-white, lip white with dark purple veins and dense purple suffusion . . . . *C. flabelliformis*

Cochleanthes aromatica (Reichenbach f.)

R. E. Schultes & Garay, *Bot. Mus. Leafl.* 18: 323. 1959. Basionym: *Zygopetalum aromaticum* Reichenbach f., *Bot. Zeit. (Berlin)* 10: 668. 1852. Type: Panama, Chiriquí volcano, *Warscewicz* (holotype: W).

SYNONYMS
Bollea wendlandiana (Reichenbach f.) Reichenbach f., *Garden & Forest* 1: 315. 1888.
Huntleya aromatica hort. ex B. S. Williams, *Orchid Growers Manual*, ed. 7: 756. 1894.
Chondrorhyncha aromatica (Reichenbach f.) P. H. Allen, *Ann. Missouri Bot. Gard.* 36: 85. 1949.
Warczewiczella aromatica Reichenbach f., *Ann. Bot. Syst.* 6: 654. 1863.

Zygopetalum wendlandi Reichenbach f., *Beitr. Orch. Centr.-Amer.*: 74. 1866.
Warczewiczella wendlandii hort. ex Nash, *Standard Cycl. Hort. (Bailey)* 6: 3506. 1917.
Warczewiczella wendlandi (Reichenbach f.) Schlechter, *Beih. Bot. Centralbl.* 36(2): 494. 1918.

DESCRIPTION The sepals and petals are white, pale green, or yellow-green, the lip is blue darkening to purple at the base, usually having a white border, and the callus is violet-blue. The sepals and petals are somewhat twisted. The lip has an isthmus basally, then widens out at mid length to the multilobulate apex, which gives the lip margin a crisped or undulate appearance. The lip disc has a basal callus that is lunate or rhombic, spreading or radiating, with many keels. The column is erect, widest at the apex, and narrowly winged.

COMMENT This is a fairly common species, and many pictures of it abound on the Internet.

MEASUREMENTS Leaves 15–30 cm long, 1.5–2.5 cm wide; inflorescence 7–9 cm long; sepals 2.5–3.0 cm long, 0.6–0.8 cm wide; petals 2.5–3 cm, 0.5–0.6 cm wide; lip 2.2–2.5 cm long, 1.5–1.7 cm wide; column 1.0 cm long.

ETYMOLOGY Latin *aromaticus*, having an aroma, referring to the scent of the flower.

DISTRIBUTION AND HABITAT Known from Costa Rica and Panama.

FLOWERING TIME August and September.

Cochleanthes flabelliformis (Swartz) R. E. Schultes & Garay, *Bot. Mus. Leafl.* 18: 324. 1959. Basionym: *Epidendrum flabelliforme* Swartz, *Prod. Veg. Ind. Occ.*: 123. 1788. Type: Jamaica, *O. Swartz s.n.* (holotype: BM); type species of *Cochleanthes*.

SYNONYMS
Cymbidium flabelliforme (Swartz) Swartz, *Nov. Act. Soc. Sc. Upsal.* 6: 73. 1799.
Zygopetalum cochleare Lindley, *Bot. Reg.* 22: plate 1857. 1836.
Cochleanthes fragrans Rafinesque, *Fl. Tellur.* 1: 45. 1836.
Zygopetalum cochleata Paxton, *Mag. Bot.* 4: 279 (Index). 1838, err. typ.
Warczewiczella cochlearis (Lindley) Reichenbach f., *Bot. Zeit. (Berlin).* 10: 714. 1852.
Huntleya imbricata hort. Hamburg ex Reichenbach f., *Bot. Zeit. (Berlin).* 10: 714. 1852, *in syn.*
Eulophia cochlearis (Lindley) Steudel, *Nomencl. Bot.*, ed. 2, 1: 605. 1857.
Warczewiczella lueddemanniana Reichenbach f., *Hamburger Garten-Blumenzeitung* 16: 179. 1860. Type: W.
Zygopetalum flabelliforme (Swartz) Reichenbach f., *Ann. Bot. Syst.* 6: 652. 1863.

Zygopetalum conchaceum Hoffmannsegg ex Reichenbach f., *Ann. Bot. Syst.* 6: 653. 1863.
Zygopetalum lueddemannianum Reichenbach f., *Ann. Bot. Syst.* 6: 653. 1863.
Cymbidium flabellifolium Swartz ex Grisebach, *Fl. Brit. W. Ind.*: 629. 1864 (*sphalm.*).
Warczewiczella cochleata (Lindley) Barbosa Rodrigues, *Struct. des Orchid*: plate 13, fig 4. 1883.
Zygopetalum gibeziae N. E. Brown, *Lindenia* 4: 79, plate 181. 1888.
Warczewiczella gibeziae (N. E. Brown) Stein, *Orchideenbuch*: 592. 1892.
Warczewiczella flabelliformis (Swartz) Cogniaux, *Symb. Antill. (Urban)* 4: 182. 1903, as *Warscewiczella*.
Cochleanthes lueddemanniana (Reichenbach f.) R. E. Schultes & Garay, *Bot. Mus. Leafl.* 18: 326. 1959.
Chondrorhyncha flabelliformis (Swartz) Alain, *Phytologia* 8: 369. 1962.

DESCRIPTION The sepals and petals are green-white, the lip is white with dark purple veins and a dense violet suffusion, the callus is purple-veined, and the column is white with some purple lines on the ventral surface. The lip incurves at either side of the callus and the lip margin is serrulate. The callus is fan-shaped near its base, with a slight overhang at mid-point, and the callus margin is wavy but not toothed. The column is arcuate, with a short column foot with a small erect spur on the ventral base. PLATE 36.

MEASUREMENTS Leaves 24 cm long, 4.5 cm wide; dorsal sepal 2.7 cm long, 1.1 cm wide; lateral sepals 3 cm long, 1.2 cm wide; petals 2.5 cm long, 0.8 cm wide; lip 4 cm long, 4 cm wide; column 1.2 cm long.

ETYMOLOGY Latin *flabelliformis*, fan-shaped, referring to either the foliage or the callus.

DISTRIBUTION AND HABITAT Known throughout the American tropics, in the West Indies, and from Mexico to Brazil, at 600 m in elevation.

FLOWERING TIME Summer.

Daiotyla Dressler, *Lankesteriana* 5(2): 92. 2005.

Daiotyla is a genus formed of necessity by the break-up of the genus *Chondro-rhyncha*. *Daiotyla* species have small flowers with rather huge callus mounds that are easily seen from a front view of the flower.

Vegetatively, the species of *Daiotyla* are similar to *Stenia* in that the leaves are obovate, being much wider apically than at the conduplicate leaf base. The small plants generally have few (two or three) leaves, with a few basal sheaths. The leaves are green without a grayish hue. The inflorescence is semierect, with one or two bracts along its length. The flowers are small but a good size relative to the amount of foliage. All the species have a thick dorsal sepal that is concave especially at the apex. The dorsal sepal has the appearance of a hook. The thin concave lateral sepals, as they reflex backward, also appear hooklike. The lateral sepals curl lengthwise on themselves and in some species in this genus form complete tubes on either side of the column foot. The petals are held parallel to the column and reflex at their apexes. The dorsal sepal and the petals cover the column, and along with the lip give the flower a bell-shaped appearance. The lip is simple, having one lobe that may be notched at the apex. The basal callus consists of two thick (0.3 cm deep), easily visible mounds that extend to the mid length of the lip. The mounds are joined basally and along most of their length, only separated apically by a notch of varying length. The column is widest at the stigma and slightly bent. The clinandrium is subapical at about a 45-degree angle from the axis of the column. The column foot is held at variable angles to the axis of the column. The four pollinia are flattened, of two different sizes, attached to a short but well-developed stipe with a shield-shaped viscidium. The rostellum is tripartite and cartilaginous with a central tooth.

Molecular data place *Daiotyla* close to *Stenia*, well removed from *Chondro-rhyncha*.

Etymology
Greek *daio*, divide, and *tyle*, knot or callus, referring to the divided callus.

List of the Species of *Daiotyla*

*Daiotyla albicans** (type species of *Daiotyla*)

*Daiotyla crassa**

Daiotyla maculata

Daiotyla xanthina

* Molecular sampling confirms placement of this species within the genus
   Daiotyla.

Key to the Species of *Daiotyla*

1a. Lip apex whole, without a notch . . . . go to 2

1b. Lip apex notched. . . . . go to 3

2a. Lip with dark maroon spots. . . . . *D. crassa*

2b. Lip buttercup-yellow with no spots, some specimens with maroon on
      callus and/or column. . . . . *D. xanthina*

3a. Lip white with two red blotches on callus, the remainder of the flower
      yellow-cream, sometimes blotches on callus lacking . . . . *D. albicans*

3b. Lip with diffuse red spots, more dense on interior of the lip, flower yellow-
      green . . . . *D. maculata*

Daiotyla albicans (Rolfe) Dressler,
Lankesteriana 5(2): 92. 2005. Basionym:
Chondrorhyncha albicans Rolfe, *Bull. Misc.
Inform. Kew*: 195. 1898. Type: Costa Rica, *W.
Rothschild s.n.* (holotype: K); type species of
Daiotyla.

DESCRIPTION The sepals and petals are cream-
white, the lip is yellow-white, the callus is yellow, and the middle of the callus mound usually has a longitudinal red blotch. Sometimes
the callus is cream-yellow without the red
blotches. The lip is one-lobed with an apical
notch, and is crisped on the lateral margins.
The callus is fleshy, made of two mounds joined
in the middle, with blunt apical margins.

COMMENT There is a good picture of this species in *Native Colombian Orchids* (R. Escobar
1994, 5: 685) labeled *Chondrorhyncha* sp. An
excellent photo of this species by Franco Pupulin can be found at www.neotrop.org/gallery
labeled *C. albicans* and another at www.jardin
botanicolankester.org.

MEASUREMENTS Leaves 8–20 cm long, 2.2–4.3
cm wide; inflorescence 6 cm long; dorsal sepal
1.6 cm long, 0.7 cm wide; lateral sepals 2 cm
long, 0.9 cm wide; petals 2 cm long, 0.9–1.1 cm
wide; lip 2.2–2.4 cm long, 2 cm wide; column 1
cm long.

ETYMOLOGY Latin *albescens*, becoming white,
referring to the flower color.

DISTRIBUTION AND HABITAT Known from
Costa Rica, Panama to Colombia, at 250–1700
m in elevation.

FLOWERING TIME March to December.

Daiotyla crassa Dressler, *Lankesteriana* 5(2):
92. 2005. Basionym: *Chondrorhyncha crassa*
Dressler, *Orchidee (Hamburg)* 34(6): 222. 1983.
Type: Panama, Chiriquí, Fortuna Valley,
north side of river near La Sierpe, 11 May
1982, *R. L. Dressler 6055* (holotype: US).

DESCRIPTION The sepals and petals are pale
yellow or white, the lip is cream-white or pale
yellow and spotted with purple, the spots
become denser laterally. The callus is flushed
with purple basally, becoming darker apically
and forming two purple spots, and the column
is yellow sometimes with red or purple ventrally. The lip is strongly concave and the mar-

gins are crisped, especially apically. The callus is obovate with an emarginate apex, which is raised prominently off the disc, the callus is thickest laterally with a weak median grove, sometimes with a few teeth lateral to the midline apex.

COMMENT There is an excellent photo of this species by Franco Pupulin at www.neotrop. org/gallery labeled *Chondrorhyncha crassa* and another at www.jardinbotanicolankester.org.

MEASUREMENTS Leaves 6.5–13 cm long, 2.3–4.0 cm wide; inflorescence 3.0–4.5 cm long; dorsal sepal 1.2–1.3 cm long, 0.6 cm wide; lateral sepals 1.6–1.8 cm long, 0.8 cm wide; petals 1.4–1.5 cm long, 0.8 cm wide; lip 1.9–2.0 cm long, 2.0–2.1 cm wide; column 0.7 cm long, 0.5 cm wide.

ETYMOLOGY Latin *crassus*, thick, referring to the callus.

DISTRIBUTION AND HABITAT Known from Costa Rica and Panama, at 800–1200 m in elevation.

FLOWERING TIME Sporadically throughout the year.

Daiotyla maculata (Garay) Dressler,
Lankesteriana 5(2): 92. 2005. Basionym: *Chondrorhyncha maculata* Garay, *Orquideología* 4: 161. 1969. Type: Colombia, Caldas, Santuario Caldas, 1800 m, *Gilberto Escobar 543* (holotype: AMES).

DESCRIPTION The flower is yellow-green and the lip has small purple spots, which are more dense basally and become sparse toward the apex of the lip. The lip is concave basally, flaring apically; the apical margins are undulate, and the apex is shallowly notched. The callus is bilobed.

COMMENT There is a good picture of this species in *Native Colombian Orchids* (R. Escobar 1994, 5: 685).

MEASUREMENTS Leaves 10 cm long, 2.5 cm wide; inflorescence 3 cm long; dorsal sepal 1.6 cm long, 0.6 cm wide; lateral sepals 2 cm long, 0.7 cm wide; petals 1.6 cm long, 0.8 cm wide; lip 2 cm long, 1.6 cm wide; column 1 cm long.

ETYMOLOGY Latin *maculatus*, spotted, referring to the spots on the lip.

DISTRIBUTION AND HABITAT Found only from Colombia, at 1800 m in elevation.

FLOWERING TIME Not recorded.

Daiotyla xanthina Dressler & Pupulin,
Orquideología 15(1): 5. 2007. Type: Colombia, flowered in cultivation at Lankester Botanical Garden in Turialba, Costa Rica, JBL-13613, 22 November 2005, *F. Pupulin 5891* (holotype: COL).

DESCRIPTION The flowers are yellow. Some specimens have a maroon medial blotch on the callus and a maroon dot on the ventral surface of the column. The lateral sepals curl lengthwise on themselves to form complete tubes with an apical hook at either side of the column foot. The petals are widest at three-quarter length and reflex at their apex. The lip has a crenulate margin with a slight notch at its apex. The callus is markedly thick (0.3 cm wide) and extends to the lateral edges of the lip. The column is widest at the stigma with slight wings at that mid column length. The clinandrium is held at a 45-degree angle with the column hooding over the clinandrium and has no apical tooth. The column foot is held at 80 degrees to the main column. PLATE 37.

MEASUREMENTS Leaves 12 cm long, 3–4 cm wide; inflorescence 3 cm long with one bract; dorsal sepal 1.5 cm long, 0.3 cm wide; lateral sepals 2.2 cm long, 0.6 cm wide; petals 1.8 cm long, 0.8 cm wide; lip 2 cm long, 1.3 cm wide; column 0.8 cm long.

ETYMOLOGY Greek *xantho*, yellow, referring to the color of the flower.

DISTRIBUTION AND HABITAT Known only from Colombia.

FLOWERING TIME October and November.

Echinorhyncha Dressler, *Lankesteriana* 5(2): 94. 2005.

Echinorhyncha is another genus formed of necessity by the breakup of the genus *Chondrorhyncha*; it groups species with a callus plate that has a central keel or keels and apical teeth, and bristly sea urchinlike appendages on the underside of the column near the stigma.

The species of *Echinorhyncha* have very short stems with few to many leaves that are long and taper at the ends, not widening much more than the foliar sheaths below. The flowers are cupped or bell-shaped. The dorsal sepal and the petals are held parallel to the column and are concave at their apexes except for *E. antonii*, which reflexes at the apexes of its sepals and petals. The lateral sepals are either reflexed and held up, or held forward and parallel to the column. The lip is fairly simple with lateral margins that incurve while the apex remains concave or reflexes slightly. The callus is a central ridge or plate with central keels and teeth. The column bends slightly and is short, and the column foot has a ventral ridge with a tooth (the base of the column is not shown on the type drawing of *E. litensis*, so I am uncertain if this species has this trait). The clinandrium is on the ventral surface of the column (the clinandrium could be said to be hooded by the column). The column bears two or more bristly sea urchinlike appendages on the underside basal to the stigma. The stigma is round and open. The stipe is minimal; the viscidium is shield-shaped, pandurate, or narrowed basally, except in *E. vollesii* which is more quadrate with a mutedly narrowed base.

Echinorhyncha is sister to the genera *Benzingia*, *Daiotyla*, *Euryblema*, *Kefersteinia*, and *Stenia*. The flower shape, callus, and lip of *Echinorhyncha* are similar to those of *Euryblema*. *Echinorhyncha* has sea urchinlike glands on the underside and the clinandrium is on the ventral surface, whereas the clinandrium of *Euryblema* is located more on the apical column and the stigma is a different shape. In addition *Euryblema* lacks the ventral tooth at the base of the column foot.

Etymology

Greek *echinos*, sea urchin or hedgehog, and *rhynchos*, beak, referring to the appendages under the column.

List of the Species of *Echinorhyncha*

Echinorhyncha antonii
Echinorhyncha ecuadorensis
*Echinorhyncha litensis** (type of genus *Echinorhyncha*)
Echinorhyncha vollesii

* Molecular sampling confirms placement of this species within the genus *Echinorhyncha*.

Key to the Species of *Echinorhyncha*

1a. Lip notched at the apex, callus extends laterally to the lip margin . . . go to 2
1b. Lip not notched at apex, callus does not extend to the lateral margins . . . .
. . . . go to 3
2a. Callus with three central keels flanked by two shorter keels that extend
to the lateral margin . . . . *E. antonii*
2b. Callus with a single central keel, seven-toothed at the apical margin
of the callus with a central sulcus free of teeth. . . . . *E. litensis*
3a. Lip margin denticulate, callus of five central keels with five apical teeth,
lip apical margin red . . . . *E. vollesii*
3b. Lip margin whole, callus of three keels, middle keel longest, lip apical
margin green-yellow. . . . .*E. ecuadorensis*

Echinorhyncha antonii (P. Ortíz) Dressler,

Lankesteriana 5(2): 94. 2005. Basionym: *Chondrorhyncha antonii* P. Ortíz, *Orquideología* 19: 14. 1994. Type: Colombia, Chocó, 28 October 1993, *P. Ortíz 1054* (holotype: UJCOL).

DESCRIPTION The sepals and petals are white, and the lip is yellow with brown-purple spots in the throat. The dorsal sepal and petals extend forward with the apex reflexing back, in contrast to the lateral sepals, which extend backwards. The lip is ovoid with a flat base and a slightly notched apex and the apical margin is undulate. The lip has a rib or keel located high at the lip base and low for the rest of its length to the apex of the callus. The callus is a central three-pointed structure flanked by two shorter lobes with external margins arching toward the margin of the disc. The column is arched with obtuse wings lateral to the stigma, and there are two hairy (bristles) glands basal to the stigma. PLATE 38.

MEASUREMENTS Leaves 50 cm long, 3 cm wide; inflorescence 8 cm long; dorsal sepal 2.2 cm long, 0.5 cm wide; lateral sepals 3 cm long, 0.5 cm wide; petals 2.6 cm long, 0.5 cm wide; lip 2.5 cm long and wide; column 1.2 cm long.

ETYMOLOGY Named to honor José Antonio González, an orchid grower in Cali, Colombia, who owned a plant.

DISTRIBUTION AND HABITAT Known only from Colombia.

FLOWERING TIME October.

Echinorhyncha ecuadorensis (Dodson)

Dressler, *Lankesteriana* 5(2): 94. 2005. Basionym: *Chondrorhyncha ecuadorensis* Dodson, *Icon. Pl. Trop.*, ser. 2, 5: plate 415. 1989. Type: Ecuador, Pichincha, near Mindo, 1400 m, *Hirtz 395* (holotype: SEL).

DESCRIPTION The flowers are pale yellow-white or yellow with a red margin on the lip and red spots on the white column. The dorsal sepal and petals are recurved and erect apically, though the dorsal sepal is concave at its apical margins. The lateral sepals are reflexed and arch upward. The lip is ovoid, trullate when spread, truncate at the apex, and the apical margins recurve slightly. The callus forms a central ridge terminating in three lobes, the middle one being longer and emarginate at the apex. The column is flattened on the underside, expanded into narrow wings above the middle, with two tufts of hairs (bristles) on each wing ventral surface. The large stigma is ovoid. PLATE 39.

COMMENT I believe that plate 117 in volume 1 of *Native Ecuadorian Orchids* (Dodson and R. Escobar 1993), labeled *Chondrorhyncha hirtzii*, is clearly misidentified; the plant shown is *Echinorhyncha ecuadorensis*.

MEASUREMENTS Leaves 30 cm long, 4 cm wide; inflorescence 10 cm long; dorsal sepal 2 cm long, 0.7 cm wide; lateral sepals 2.5 cm long, 1 cm wide; petals 2 cm long, 0.9 cm wide; lip 4 cm long and wide; column 1.5 cm long.

ETYMOLOGY Named for the country of its origin.

DISTRIBUTION AND HABITAT Known only to Ecuador, at 1400 m in elevation, in very wet montane cloud forest.

FLOWERING TIME Most of the year.

Echinorhyncha litensis (Dodson) Dressler,

Lankesteriana 5(2): 94. 2005. Basionym: *Chondrorhyncha litensis* Dodson, *Icon. Pl. Trop.*, ser. 2, 5: plate 417. 1989. Type: Ecuador, Esmeraldas, 850 m, 7 July 1988, *Dodson &*

Gentry 17570 (holotype: QCNE; isotype: RPSC); type species of *Echinorhyncha*.

DESCRIPTION The sepals, petals, and lip are yellow-white and the lip is variably blotched with red. The dorsal sepal and petals recurve at their apexes. The lateral sepals spread to the sides and forward, parallel to the column. The lip is obovate, bilobed (notched) at the apex, and the lateral margins are erose. The lip is minimally flared for its apical quarter. The callus is broad and raised with seven apical teeth and a central keel that is thicker at the lip base, lower in the middle, and terminates in the central tooth. The apical margin of the callus is erose. The column is terete, flattened on the underside, with small ovate wings and pair of small pubescent protuberances (bristles) on each side of the stigma. PLATE 40.

MEASUREMENTS Leaves 22 cm long, 3 cm wide; inflorescence 12 cm long; dorsal sepal 3 cm long, 0.9 cm wide; lateral sepals 3.2 cm long, 1.1 cm wide; petals 3 cm long, 0.9 cm wide; lip 3.5 cm long, 3 cm wide; column 2 cm long.

ETYMOLOGY Named for Lita, the nearest population center to the type locality.

DISTRIBUTION AND HABITAT Known from a small area on the border of Colombia and Ecuador, at 850–1400 m in elevation, in a montane heavy rain forest.

FLOWERING TIME Most of the year.

Echinorhyncha vollesii (G. Gerlach,

Neudecker & Seeger) Dressler, *Lankesteriana* 5(2): 94. 2005. Basionym: *Chondrorhyncha vollesii* G. Gerlach, Neudecker & Seeger, *Orchidee (Hamburg)* 40(4): 131. 1989. Type: Colombia: Nariño, La Planada, 1800 m (holotype: HEID).

DESCRIPTION The sepals and petals are pale yellow; the lip is spotted lightly on its interior with red-pink spots and with indistinct blotches of red-pink on the apical lip margin. The dorsal sepal and petals recurve at their apexes, the dorsal sepal is semierect and the lateral sepals spread back and up. The lip is rhomboid with denticulate margins and a short reflexed portion at the apical margin. The callus is slightly raised with five keels producing five dull teeth apically. The column is terete, flattened on the underside, with small ovate wings and a pair of pubescent protuberances below each side of the stigma. The entire column is slightly pubescent. PLATE 41.

COMMENT The picture in *Native Colombian Orchids* (R. Escobar 1994, 5: 685) shows concave petals and dorsal sepal. I wonder if the flower is just immature, or if the flower is misidentified. The picture accompanying the original description shows reflexing dorsal sepal and petals. The plant pictured in the present volume as Plate 41 has a slightly concave dorsal sepal but otherwise fits this species. *Echinorhyncha vollesii*

and *E. ecuadorensis* are very similar, differing in the number of keels in the callus and the ventral side of the column, but their general shape is the same.

MEASUREMENTS Leaves 15 cm long, 1.5 cm wide; inflorescence 10 cm long; dorsal sepal 1.6 cm long, 0.7 cm wide; lateral sepals 2.1 cm long, 0.9 cm wide; petals 1.6 cm long, 0.7 cm wide; lip 2 cm long, 2.5 cm wide; column 1.2 cm long.

ETYMOLOGY Named for Hans Volles, honoring his knowledge of Colombian and Ecuadorian orchids.

DISTRIBUTION AND HABITAT Known only from Colombia and Ecuador, at 1800 m, in cloud forest with species of *Dracula* and *Masdevallia*.

FLOWERING TIME Not recorded.

Euryblema Dressler, *Lankesteriana* 5(2): 94. 2005.

Euryblema is another genus created as result of molecular data that excluded certain species from the genus *Chondrorhyncha sensu stricto*. In doing so, Dressler makes a small but easily defined genus.

Euryblema species are relatively large tuft-forming epiphytes without pseudobulbs. The leaves are oblanceolate, narrowed basally, not widening much more than the basal sheath, and they are pointed apically. Plants of this genus are easily recognized by the red-spotted leaf sheaths (or leaf bases). The dorsal sepal, though being somewhat parallel with the column, is also somewhat erect, and the ventral surface of the dorsal sepal is showy. The lateral sepals reflex back and up. The petals are held parallel with the column covering the column from above, and the petals reflex at their apexes. The lip and the petals form a tube. The lip is concave at the base and flares apically. The lip has a very short chin or claw at its base. The callus is broad and keeled with not really a thickening but a plate or shelf covering the basal half of the lip. The column is more or less straight, with lateral wings and a ventral keel that does not extend to the base of the column. The stigma is small and wedgelike. The clinandrium angles back 20 degrees from the line of the column so that it is not really hooded by the column, but is not all apical. The viscidium is broad and shield- or trowel-shaped; the stipe is minimal.

Euryblema is sister to the genera *Benzingia*, *Daiotyla*, *Echinorhyncha*, *Kefersteinia*, and *Stenia*. The flower shape of *Echinorhyncha* resembles that of *Euryblema*, having very similar callus and lip structures. *Echinorhyncha* species have sea urchinlike glands on the underside and the clinandrium is on the ventral surface, whereas *Euryblema* species have the clinandrium more on the apical column and the stigma is a different shape. *Euryblema* species also lack the ventral tooth at the base of the column foot.

Etymology
Greek *eurys*, broad, and *blema*, blanket or cover, referring to the callus shape.

List of the Species of *Euryblema*

*Euryblema anatonum** (type species of *Euryblema*)

*Euryblema andreae**

* Molecular sampling confirms placement of this species within the genus
  Euryblema.

Key to the Species of *Euryblema*

1a. Lip 4 cm long, flower yellow, with red spots basally on lip . . . . . *E. anatonum*

1b. Lip 5 cm long, flower yellow-white, densely covered with red spots
      especially the apical portions of the flower parts . . . . *E. andreae*

Euryblema anatonum (Dressler) Dressler,
Lankesteriana 5(2): 94. 2005. Basionym:
Cochleanthes anatona Dressler, *Orchidee
(Hamburg)* 34(4): 160. 1983. Type: Panama,
Coclé, near Aserradero El Copé, ca. 8 km
northwest of El Copé, 750–850 m, very
wet cloud forest, 1 September 1977, *R. L.
Dressler 5690* (holotype: US); type species of
Euryblema.

SYNONYM

Chondrorhyncha anatona (Dressler) Senghas,
Orchidee (Hamburg) 41(3): 91. 1990.

DESCRIPTION The sepals and petals are pale
yellow-green, and the lip is yellow lightly speck-
led or suffused with red or red-brown basally.
The lip is rhombic-ovate, weakly three-lobed,
and the apical margin is strongly crisped. The
callus reaches mid length of the lip blade, is
flat, erose, and pilose basally, with a central
midline keel. The column is winged with a ventral keel and has short hairs at the ventral base.
PLATE 42.

MEASUREMENTS Leaves 15–36 cm long, 3.4–4.5
cm wide; inflorescence 6.0–7.5 cm long; dorsal sepal 2.7–3.2 cm long, 0.8–0.9 cm wide; lateral sepals 3.4–4.2 cm long, 0.8 cm wide; petals
3.6–3.8 cm long, 0.9–1.0 cm wide; lip 4.0–4.2
cm long, 3 cm wide; column 2.2 cm long, 0.5
cm wide.

ETYMOLOGY Greek *an*, if, and *atona*, languidly,
referring to how the sepals are held.

DISTRIBUTION AND HABITAT Known from
Panama, at 750–850 m in elevation, in very wet
cloud forest.

FLOWERING TIME September.

Euryblema andreae (P. Ortíz) Dressler,
Lankesteriana 5(2): 94. 2005. Basionym:
Chondrorhyncha andreae P. Ortíz, *Orquideología*
19: 13. 1994. Type: Colombia, Chocó, Bajo
Río Atrato, flowered in cultivation of
Senén Rendón, October 1993, *P. Ortíz 1055*
(holotype: UJCOL).

DESCRIPTION The dorsal sepal and petals are
yellow-green to yellow-white, becoming more
green basally, with a pink suffusion on the apical upcurved portion; the petals have red or
red-purple spots on the upcurved apical portion. The lateral sepals are yellow-green to
white, becoming more green basally. The lip is
white with maroon spots becoming confluent
on the apical margin. The callus is yellow with
the underside (outside of flower) spotted with
maroon. The column is white-cream with red
spots, and the anther is white. The dorsal sepal
and petals form a hood over the column, flexing upward at the apical end of the column.
The lateral sepals reflex back and their lower
margins are concave and inrolled. The lip is
three-lobed; the lateral lobes curl upward to the
column, the midlobe flexes down at mid length
only slightly and flares to form a skirt, the lip

margins are undulate and mildly crispate. The callus is more like a disc or sheet over the concave portion of the lip with a set of three keels; the lateral two keels are barely discernable: the mid keel is more prominent, hirsute, and runs down the midline to end in two teeth; the rest of the callus plate apex is a slight ridge that extends to the lateral edges of the lip. The column is minimally winged with a ventral keel and has short hairs at the ventral base. The flower has the fragrance of limes. PLATE 43.

MEASUREMENTS Leaves 50–60 cm long, 6 cm wide; inflorescence 8–10 cm long; dorsal sepal 3.8 cm long, 0.8 cm wide; lateral sepals 5 cm long, 0.8 cm wide; petals 3.3 cm long, 1 cm wide; lip 5 cm long, 3.8 cm wide; column 1.5 cm long.

ETYMOLOGY Named for Andrea Niessen de Uribe, who cultivated the plant.

DISTRIBUTION AND HABITAT Known only from Colombia (and perhaps Ecuador).

FLOWERING TIME October.

Hoehneella Ruschi, *Publ. Arq. Publico Estad. Esp. Santo*: 3. 1945. Reprinted as *Orquid. Nov. Estad. Esp. Santo.*

Hoehneella was established in 1945 by Brazilian taxonomist Augusto Ruschi, who separated *Zygopetalum* species with pseudobulbs from *Zygopetalum* species without pseudobulbs. Garay (1969) combined *Chaubardia* and *Hoehneella*, making them congeneric, because they both have pseudobulbs. Later, Garay (1973) accepted *Hoehneella* as distinct and pointed out that the lateral sepals of *Hoehneella* are not gibbose at the base and that the lip is sessile and articulates with the column foot. He stated that this was unlike *Chaubardia*, which has lateral sepals that are gibbose at the base and a lip that is unguiculate. He pointed out that *Hoehneella* species have a column that is not winged and an anther that is crested in contrast to the column of *Chaubardia*, which has wings and a noncrested anther.

Hoehneella species also differ from *Chaubardia* in lacking the doubling back, or reflection, at the base of the lip. Both genera have callus teeth that are merely raised, not long and fimbriate. The two share the minimal or absent stipe on the pollinia, and the ovate stigma and the hooded clinandrium of the column.

Hoehneella species possess small pseudobulbs similar to *Chaubardia* species. The leaves have basal sheaths and the conduplicate leaf articulates with the sheath, widening at mid length and then tapering to the pointed apex. The sepals and petals spread so the flower is fairly open. The lip sticks out horizontally in front. The lip base is simple and has no claw. The callus consists of several keels extending onto the midlobe, and the lip is pubescent. The column is straight with no true wings. The clinandrium is hooded by the column apex. The viscidium is transversely elliptic and lacks a stipe.

The inclusion or exclusion of *Hoehneella* species in the genus *Chaubardia* has not been determined using molecular studies because of lack of material. *Hoehneella* species have not been collected for many years. Brazilian taxonomists have told me that they don't really think the species are extinct, just that they haven't been seen and reported in many years. These individuals feel it's just a matter of going out and purposely looking for the plants.

Etymology

Named for Federico Carlos Hoehne, a professor who worked on the taxonomy of the Brazilian orchids.

List of the Species of *Hoehneella*

Hoehneella gehrtiana (type species of *Hoehneella*)
Hoehneella heloisae

Key to the Species of *Hoehneella*

1a. Callus widest at base, sepals and petals green . . . .*H. gehrtiana*
1b. Callus thin basally, at mid length widening to become fan-shaped, sepals and petals yellow. . . . . *H. heloisae*

Hoehneella gehrtiana (Hoehne) Ruschi, *Orquid. Nov. Estad. Esp. Santo*: 5. 1945; cf. *Gray Herb. Card Cat.* (*sphalm*. Gehrtii). Basionym: *Warczewiczella gehrtiana* Hoehne, *Arq. Bot. Estad. São Paulo*, n.s., f.m., 1: 21. 1938. Type: Brazil, São Paulo, Pirajussara, 1933–1934, *A. Gehrt s.n.* (holotype: SP); type species of *Hoehneella*.

SYNONYM
Chaubardia gehrtiana (Hoehne) Garay, *Orquideología* 4: 143. 1969.

DESCRIPTION The flower is a clear green with a white lip and the column is green with red spots on the ventral surface. The lip is trilobed, the small lateral lobes are erect, and the mid-lobe margin is mildly undulate. The lip disc, except the callus, is minutely pubescent. The callus is basal, broadest at its base, with five to nine ribs or keels and having five teeth on its apical margin. The column is straight, slightly widening lateral to the stigma, and there are no wings on the column. The clinandrium is on the ventral surface; the apex of the column hoods over the anther and terminates in a small tooth. The plants often produce cleistogamous flowers. PLATE 44.

COMMENT There are photographs of the flower and plants, and an excellent drawing of the flower parts by Karl Senghas in *Die Orchidee* (1993, 44 [2]).

MEASUREMENTS Leaves 7–15 cm long, 1.5–2.2 cm wide; inflorescence 5–7 cm long; sepals and petals 1.5–1.8 cm long, 0.5–0.6 cm wide; lip 1.5 cm long, 1.1 cm wide; column 0.8 cm long.

ETYMOLOGY Named for Augusto Gehrt, who discovered the species and preserved the type specimen.

DISTRIBUTION AND HABITAT Known from Brazil in the regions of São Paulo, Serra do Mar, Rio de Janeiro, Espírito Santo, and in the interior of Brazil, at 700–800 m in elevation.

FLOWERING TIME Not recorded.

Hoehneella heloisae Ruschi, *Orquid. Nov. Estad. Esp. Santo*: 4. 1945. Type: Brazil, Espírito Santo, Santa Tereza, 16 November 1939, *A. Ruschi s.n.* (holotype: MBNL).

SYNONYM
Chaubardia heloisae (Ruschi) Garay, *Orquideología* 4: 143. 1969, as *C. heliosae*.

DESCRIPTION The flower is green-yellow with maroon warts on the lip and face of the column. The lip is trilobed, the lateral lobes are much smaller than the middle lobe, and the middle lobe is undulate and pubescent internally. The basal callus is one narrow keel with a thickened base that ends with a fan of seven keels. The most lateral keels seem to make a connection with the base of the lateral lobes of the lip, but do not extend onto the lateral lobes; the apical callus edges extend over to the median lip lobe. There are no wings on the column.

MEASUREMENTS Leaves 4–15 cm long, 1.5–2.5 cm wide; inflorescence 3.0–6.5 cm long; dorsal sepal 1.5 cm long, ca. 0.5 cm wide; lateral sepals 1.7–1.8 cm long, ca. 0.6 cm wide; petals 1.5 cm long, ca. 0.5 cm wide; lip 1.5 cm long, 1.1 cm wide; column 0.9 cm long.

ETYMOLOGY Named for Heloisa Torres, director of the National Museum of Rio de Janeiro.

DISTRIBUTION AND HABITAT Known from the Espírito Santo region of Brazil, but has not been seen for many years.

FLOWERING TIME November.

Huntleya Bateman ex Lindley, *Bot. Reg.* 23: plate 1991. 1837.

The genus *Huntleya* was described in 1837 by John Lindley, who validated the previously unpublished genus *Huntleya* of Mr. Bateman. At that time two species were described, *H. meleagris* and the less common *H. sessiliflora*. The two species (possibly synonyms) are physically alike, varying mostly by color.

The flowers of *Huntleya* are striking and very desirable; however, the plants are somewhat difficult to grow, especially out of their tropical climates. Finding blooming plants in northern latitudes was rare, so familiarity with *Huntleya* was mostly by paintings. New specimens were labeled with known species names and not renamed when new determinations of the species were made. *Huntleya burtii* was described in 1872, thirty-five years after *H. meleagris*, under the genus *Batemannia*, even though examples of the species had been found long before that date. *Batemannia burtii* was moved to *Huntleya* in 1889, and still only three species comprised the genus: *H. burtii*, *H. meleagris*, and *H. sessiliflora*. Even Heinrich Gustav Reichenbach, when he first described *H. burtii*, thought it could be a synonym of *H. meleagris*. As you go through magazines and books, you can see plants that appear to be *H. burtii* labeled as *H. meleagris*. The use of either name seems to depend on each author's own definition of both species.

Since these early species were recognized, other species have been found that are morphologically quite distinct. Knowing there are distinct species in the genus makes it worthwhile to try to sort out the *Huntleya burtii/meleagris* question. Either *H. meleagris* and *H. burtii* are synonyms, just widely variable, or they are distinct species that are just difficult to tell apart except for size and color. Pedro Ortíz (pers.comm.), a well-respected taxonomist in Colombia, has only seen *H. burtii* in Colombia and western South America and has never seen a native Colombian species which would be *H. meleagris* without the purple spots, although he points out that since the nonspotted *H. meleagris* is reported as being from Venezuela, it could be possible they would make it to Colombia. Ortíz says that any other physical difference between the two would be difficult to discern. Franco Pupulin confirms that the Central American plants he has seen are all *H. burtii* with the maroon markings. I have chosen to leave *H. burtii* and *H. meleagris* as separate species with *H. burtii* found more westerly and *H. meleagris* found in more eastern South America and along the Brazilian coast.

The actual plants of *Huntleya* are large for this clade. They grow as epiphytic tufts or as successive growths that ramble along the rhizome depending on the species. The rhizome is thick and obvious between the individual growths. The growths are without pseudobulbs, but have a definite thick stem, which is an extension of the rhizome. The leaves are plicate at the base, widening and lying flat apically, minimally narrowed basally to contract with the conduplicate petioles, and distichously arranged in the form of an open fan.

The inflorescence, produced from the axils of the central leaves, is strongly erect, holding the single star-shaped flower in the uppermost portions of the foliage. The flowers are flat, with the exception of the lip, which may be in the same plane as the sepals and petals or at variable angles to them. The floral parts are thick and fleshy with a very waxy texture, and some species exhibit a "quilting" pattern by elevating the area between the tessellated veins. The sepals are minimally concave and the dorsal sepal is erect and free; the lateral sepals are connate at the base and obliquely inserted into the column foot. The petals are subequal to the sepals in most species.

The lip is rather fleshy and one- to three-lobed. The lip is contracted at the base into a conspicuous claw, which forms a geniculate segment with the foot of the column. The lateral lobes of the lip are small and inconspicuous. The hypochil of the lip has a basal callus with an erect, usually semicircular, U- or V-shaped, fimbriate margin.

The column has a dilated apex with a dorsal keel; the lateral margins of the column are winged and remain definite ridges joining at the apex, forming a conspicuous thin hood over the clinandrium. The column foot is upturned, terminating in a suberect ligule (tooth) that is on top of the area of articulation with the lip. The anther is terminal, operculate, incumbent, and one-celled (or two imperfect cells). The four pollinia are subequal and waxy. The spade- or trowel-shaped viscidium is prominent, and the stipe is well-developed.

The flowers of some species are fragrant. *Huntleya* species are distributed from Nicaragua to Bolivia and Brazil.

Etymology
Named for J. T. Huntley, an ardent collector of orchidaceous plants.

List of the Species of *Huntleya*

Huntleya albidofulva
Huntleya apiculata*
Huntleya brevis
Huntleya burtii
Huntleya caroli
Huntleya citrina
Huntleya fasciata

Huntleya gustavi*
Huntleya lucida
Huntleya meleagris (type species of *Huntleya*)
Huntleya sessiliflora
Huntleya vargasii
Huntleya wallisii*

* Molecular sampling confirms placement of this species within the genus *Huntleya*.

Key to the Species of *Huntleya*

1a. Lip with one set of fimbriae on hypochil, second set below primary set and angled back . . . . go to 2
1b. Lip with one set of fimbriae on hypochil . . . . go to 6
2a. Second set of fimbriae form a skirt around callus, or at least a red line at juncture of hypochil and epichil . . . . *H. wallisii*
2b. Second set of fimbriae only on posterior callus, fimbriae point toward column foot, no red line below callus . . . . go to 3
3a. Lip held erect and parallel to column . . . . *H. vargasii*
3b. Lip held in same plane as the sepals and petals. . . . . go to 4
4a. Column wings rhomboid, apical lateral wing margins of column erose, maroon patch at base of petals. . . . . *H. burtii*
4b. Column wings triangular, apical lateral wing margins scalloped, no maroon patch at base of petals. . . . . go to 5
5a. Flower inflorescence very short. . . . .*H. sessiliflora*
5b. Flower inflorescence obvious. . . . . *H. meleagris*
6a. Flowers small, sepals no more than 3 cm long, lip 2 cm long. . . . . . . go to 7
6b. Flowers larger . . . . go to 10
7a. Sepals and petals more than twice as long as wide, flower yellow with brown markings on sepals and petals. . . . .*H. brevis*
7b. Sepals and petals not twice as long as wide . . . . go to 8
8a. Flower white . . . . *H. apiculata*
8b. Flower basically yellow . . . . go to 9
9a. Sepals and petals with brown-red apexes, lip red at apex . . . . *H. caroli*
9b. Lateral sepals sometime with slight red blush, lip yellow . . . . *H. citrina*
10a. Sepals and petals not twice as long as wide . . . . *H. gustavi*
10b. Sepals and petals more than twice as long as wide. . . . . go to 11
11a. Sepals and petals obovate, widening to half as wide as long . . .*H. albidofulva*
11b. Sepals and petals widest at base, quickly narrowing, never becoming half as wide as long . . . . go to 12

12a. Fimbriae on apical portion of crest much shorter than lateral fimbriae,
   lip small, 1.8–2.3 cm long, 1.2 cm wide.........................*H. lucida*

12b. Fimbriae on apical portion of crest same length as lateral fimbriae,
   lip larger, 4–5 cm long, 1–1.2 cm wide.........................*H. fasciata*

Huntleya albidofulva Lemaire, *Ill. Hort.*

15: plate 556. 1868. Type: Brazil (lectotype designated here).

DESCRIPTION The sepals and petals are white basally and brown for the apical two-thirds with two yellow windows on each side of the central vein. The lip is white, the apical tip is purple-red. The column is white with yellow on the ventral margins and green on the hooded clinandrium. The sepals and petals are obovate and acuminate. The lip hypochil has a single set of fimbriae arranged in a semicircle and all of the same length. The lip epichil is obscurely three-lobed. The column wings are large and outspreading.

COMMENT *Huntleya albidofulva* has many characteristics that make it seem to fit within the characteristics of other species. At first glance it looks like *H. meleagris*, *H. sessiliflora*, or *H. burtii*, but it has only one set of fimbriae on the hypochil.

Huntleya albidofulva, originally reported as found in Brazil and Trinidad, was reported as found again in Colombia (Garay 1969). Commenting on the Colombian specimen, Garay stated that it might be a variety of *H. burtii* or *H. wallisii*. The picture of the Colombian specimen in *Orquideología* (1969, 4: 160) shows a very pale orange-yellow flower with a white column whose shape is somewhat close to *H. sessiliflora*, but closer to *H. burtii*. The Colombian form of *H. albidofulva* has a 10-cm wide flower which is more like *H. burtii* than *H. wallisii*, but the specimen was found in Colombia in areas where *H. wallisii* is found. From the picture I can't tell if there is a second set of fimbriae on the epichil, which would help determine if the flower in the picture is indeed *H. albidofulva* or an albanistic *H. burtii*.

The original name of this species was written with a hyphen, *Huntleya albido-fulva*. The name is correct with or without the hyphen, the latter form being more modern.

MEASUREMENTS Not recorded.

ETYMOLOGY Latin *albidus*, dull white, and *fulvis*, yellow-brown, referring to the color of the sepals and petals or perhaps the column edges.

DISTRIBUTION AND HABITAT Reported to be found in the past in Brazil, Trinidad, and Colombia.

FLOWERING TIME Not recorded.

Huntleya apiculata (Reichenbach f.)

Rolfe, *Orch. Rev.* 24: 236. 1916. Basionym: *Batemannia apiculata* Reichenbach f., *Linnaea* 41: 109. 1876. Type: Colombia, 300 m (holotype: W).

DESCRIPTION The original description says the flower is pure white. The petals are widest at the basal third to half of the length. The hypochil is semiovate, the epichil is ovate. The column has two divergent column wings lateral to the stigma and an erose apex. As the plant grows, it climbs up its host.

COMMENT The term *apiculato* in the original description could be applied to the structure at the base of the lip, the sepals, or the anther hood. The drawing and photograph of the type specimen show that perhaps the petals could be thought of as apiculate, but a better description would be acute.

This is a small-flowered species for *Huntleya*, since the flowers are half as large as *H. gustavi*. This species may indeed be an earlier name for *H. citrina* and *H. waldvogelii*. The physical characteristics taken from drawings of the type specimens are very similar. If after further investigation it is demonstrated that these latter two species are synonyms of *H. apiculata*, the name *H. apiculata* would have priority.

Many pictures I have seen labeled *Huntleya apiculata* are of a flower with acuminate sepals and petals. Besides those sent to me by friends, pictures labeled *H. apiculata* are in *Native Colombian Orchids* (R. Escobar 1994, 2: 216), and *Orchid Digest* (1974, 38: 117). The flowers in these pictures have petals that are white on their lowest (toward the ground) side and basally, becoming yellow with two red-brown bands on each side of the central vein and then red-brown at the apexes. The sepals are similarly patterned but lack the white. The lip is white with red over the apical half. The column is large for a *Huntleya*. There are fine red spots on the ventral surface and the column hood is edged in a prominent green. I am unable to determine if there is one set of fimbriae on the hypochil or what the size of the flower is. My feeling is that the plants in these pictures are a variety of *H. meleagris*, or *H. albidofulva*, or a species that needs its own name, but they are not *H. apiculata*.

The *Huntleya apiculata* pictured in *Orquídeas de la Serranía del Baudó* (Urreta 2005), which depicts orchids from the Chocó, shows a flower with a white column, a few red striations on the central surface, and a white anther hood. The drawing of the flower from Chocó, Colombia, clearly shows only one set of fimbriae and fits the description, measurements, and color patterns of *H. lucida*.

MEASUREMENTS Taken from a photograph of the type specimen: Dorsal sepal 1.3 cm long, 0.5 cm wide; lateral sepals 2 cm long, 0.8 cm wide; petals 1.8 cm long, 0.5 cm wide; lip 1.5 cm long, 1.2 cm wide.

ETYMOLOGY Latin *apiculatus*, ending abruptly in a small point, referring to the point at the base of the lip, the anther hood, and in the text of the original description, the sepals.

DISTRIBUTION AND HABITAT Known only from Colombia, in the western Cordillera, at 300 m elevation, with *Warczewiczella* species.

FLOWERING TIME Not recorded.

Huntleya brevis Schlechter, *Repert. Spec. Nov. Regni Veg. Beih.* 27: 86. 1924. Type: Colombia, Barbacoas on the west coast, 250 m, July 1921, *W. Hopp 191* (holotype: B, destroyed).

DESCRIPTION The flower is yellow with brown markings and brown callus, small, and inconspicuous. The sepals are elliptic and pointed. The petals are ovate-elliptic and pointed at the apex. There is no quilting on the sepals or petals. The lip is slender and pointed. The basal crest is semilunar. The column is slightly curved and is dilated at the apex.

MEASUREMENTS Leaves 12–23 cm long, 2.0–2.8 cm wide; inflorescence 3–4 cm long; sepals and petals 2.75–3.00 cm long, 0.5 cm wide; lip 2 cm long, 1.2 cm wide; column 1.1 cm long.

ETYMOLOGY Latin *brevis*, of small extent, in reference to the inflorescence size.

DISTRIBUTION AND HABITAT Known only from Colombia, at 250 m elevation.

FLOWERING TIME Recorded as blooming in July.

Huntleya burtii (Endrés & Reichenbach f.) Pfitzer, *Die Natürlichen Pflanzenfamilien* 2(6): 205. 1889. Basionym: *Batemannia burtii* Endrés & Reichenbach f., *Gardener's Chronicle & Agricultural Gazette*: 1099. 1872. Type: Costa Rica, *Endrés s.n.* (holotype: W).

SYNONYM
Zygopetalum burtii (Endrés & Reichenbach f.) Bentham & J. D. Hooker ex Hemsley, *Biol. Cent. Amer.* 3(16): 251. 1884.

DESCRIPTION The sepals and petals are glossy dark orange-brown on the apical half, the middle portion is yellow-brown fading to yellow, and the basal quarter is ivory white. The petals have a dark red-brown to maroon blotch at the base. The lip epichil is dull brown, becoming ivory white at the lip hypochil; the crest fimbriae are purple, and the second line of fimbriae below the plate is white. The column is white dorsally, although some specimens have fine purple striations on the ventral surface. The apical crest of the column and the border of the column hood are green or white, the column wings are pale yellow-green or white, and often there is a purple spot lateral to the stigma. The anther is light yellow. The sepals and petals have heavy quilting, which appears as elliptical swellings arranged in chains along the length of the segments. The lip is pubescent on both sides and three-lobed; the midlobe is clawed and acute. The callus crest is semicircular and the upper surface has erect fimbriate appendages with a second lower set of fimbriae that faces downward and basally toward the column foot. The column is stout, fleshy, with large rhombic wings that are toothed at their margins. PLATE 45.

COMMENT *Huntleya burtii* is easy to differentiate from *H. meleagris* in that the latter lacks the dark stains at the base of the petals. The two species are also very similar and may indeed not be distinct species but rather a complex, as there are many specimens that seem to be intermediates in size and coloration. The measurements following are of a "typical" *H. burtii*.

MEASUREMENTS Leaves 20–35 cm long, 25 cm wide; inflorescence 12 cm long; sepals 4–5 cm long, 2.3–2.5 cm long; petals 4–5 cm long, 2.5 cm wide; lip 2–3 cm long, 2 cm wide; column 2.3 cm long, 1 cm wide.

ETYMOLOGY Named for Mr. Burt, a friend of James Bateman.

DISTRIBUTION AND HABITAT Known from Central America to Bolivia and Peru, at 350–1200 m in elevation, in wet montane forest.

FLOWERING TIME May to September.

Huntleya caroli P. Ortíz, *Orquideología* 23(1): 26. 2004. Type: Colombia, Nariño, Tumaco, collected by Julio César Miranda, flowered in Bogotá by Carlos Uribe Vélez, November 2002, *P. Ortíz 1140* (holotype: HPU).

DESCRIPTION The flowers are pale yellow with red-brown suffusion toward the apexes. The lip hypochil is white along with the most basal portion of the epichil, the apical epichil is red-brown, and the fimbriate crest is red-purple. There is no or very minimal quilting to the sepal or petal surfaces. The lip epichil is elliptic and convex with an acute apex, the hypochil is concave with the walls ending in acute filaments. There are seven fimbriae on each side of the callus bending towards the center and nine in the frontal center callus which lie flat, directed towards the apex of the lip. The column is somewhat curved with lateral rounded wings. PLATE 46.

COMMENT This species differs from *Huntleya gustavi* mainly by the three series of filaments in the callus crest, but also in size, shape, and ratio of the flower parts as well as color.

MEASUREMENTS Leaves 33 cm long, 4 cm wide; inflorescence 8 cm long; dorsal sepal 2.2 cm long, 1.2 cm wide; lateral sepals 2.5 cm long, 1.2 cm wide; petals 2.2 cm long, 1.2 cm wide; lip 2.5 cm long, 1.5 cm wide; column 1.3 cm long.

ETYMOLOGY Named in honor of Carlos Uribe Vélez, who has successfully grown this and other species of *Huntleya*. The name is dedicated to him, with no explanation why the name is *H. caroli*.

DISTRIBUTION AND HABITAT Known from Colombia and northern Ecuador, elevation not recorded.

FLOWERING TIME November.

Huntleya citrina Rolfe, *Bot. Mag.* 142: plate 8689. 1916. Type: Country of origin unknown, flowered in cultivation (holotype: K).

SYNONYM
Huntleya waldvogelii Jenny, *Orchidee (Hamburg)* 35(4): 131. 1984. Type: Colombia, Magdalena, Junín, flowered in cultivation, 15 April 1984, *Jenny 42* (holotype: G).

DESCRIPTION The sepals, petals, and lip are lemon-yellow, the lateral sepals sometimes have a brown patch, the lip crest and basal veins on the lip are red-brown, and the column is green-yellow with red spots on the ventral surface. The lip has a narrow base and is three-lobed. The callus is heart-shaped, forming two rows of fimbriate crests. The column is widest at mid length with wings. The column apex is crested and has an irregular lateral edge apically with two large diverging wings. PLATE 47.

COMMENT The original plant was grown by Sir Trevor Lawrence. When he died, the plant went to the Royal Botanic Gardens Kew labeled as a *Chondrorhyncha*, without any clue as to its origin. The plant is thought to have been collected by F. C. Lehmann, based on drawings and specimens, but this is not certain. Another plant bloomed in cultivation in 1984, also without specific collection location, but this plant was at least thought to be from the Junín area of Colombia, and was first thought to be the new species described as *H. waldvogelii*. The species has since been recollected in Colombia and is in cultivation. *Huntleya waldvogelii* is listed as an accepted species by the Kew World Checklist of Monocotyledons.

MEASUREMENTS Leaves 16–20 cm long, 3.5–4.0 cm wide; inflorescence 2–5 cm long; sepals 2.0–2.5 cm long, 0.8–1.0 cm wide; petals 2.2–2.8 cm long, 0.8–1.2 cm wide; lip 1.9 cm long, 0.5 cm wide; column 1.2 cm long.

ETYMOLOGY Latin *citrinus*, lemon-yellow, referring to the flower color.

DISTRIBUTION AND HABITAT Known only from Colombia, in the Department of Nariño, on mountain slopes facing the Pacific Ocean.

FLOWERING TIME April and May.

Huntleya fasciata Fowlie, *Orch. Digest* 30: 281. 1966. Type: Panama, 1000 m, September 1964, *Edna Jackson s.n.* (holotype: UCLA).

DESCRIPTION The sepals and petals are yellow with two bars of red-brown, the medial bands are often more brown than red, the apical tips are yellow. A prominent transverse band of yellow separates the brightly red-brown barred zones, and the basal portions of the sepals and petals are white. Colors of sepals and petals, starting at the base, are arranged thus: white, yellow-tan, a red-brown band, a 2-mm transverse band of yellow, then orange-red, and then yellow at the apex. The lip hypochil is white, the fimbriae and callus are white, and the epichil is dark red with a narrow yellow border. The column is white with red striations on the ventral surface. The callus crest is in the form of a semicircle.

COMMENT Fowlie (1966) says the difference between *Huntleya fasciata* and *H. lucida* is that the apex of the labellum of *H. fasciata* is recurved, whereas the labellum of *H. lucida* is abruptly upturned. In addition, he states that *H. fasciata* has petals that are narrower than the dorsal sepal, whereas *H. lucida* has petals that are broader than the dorsal sepal. In the original description the length of the inflorescence is listed at 3–4 cm, but the drawing along with the description shows an inflorescence that is at least half to three-quarters as long as the leaves. Dodson (1980) speculated that *H. fasciata* is just a synonym of *H. lucida* as there are intermediates of the two species. There is a picture of *H. fasciata* with the type description.

MEASUREMENTS Leaves 17–25 cm long, 3.8–4.0 cm wide; inflorescence 3–20 cm long; sepals and petals 3.5–4.0 cm long, 1.2–1.4 cm wide; lip 4–5 cm long, 1.0–1.2 cm wide; column 1.2–1.4 cm long.

ETYMOLOGY Latin *fasciatus*, growing in bundles, referring to the vegetative growth.

DISTRIBUTION AND HABITAT Known from Colombia and northwestern Ecuador to Honduras and Belize, from close to sea level to 1200 m in elevation.

FLOWERING TIME September.

Huntleya gustavi (Reichenbach f.) Rolfe, *Orch. Rev.* 24: 236. 1916. Basionym: *Batemannia gustavi* Reichenbach f., *Linnaea* 41: 108. 1877. Type: Colombia, 1500–1800 m, *G. Wallis s.n.* (holotype: W).

DESCRIPTION The flowers are light yellow, and, the sepals and petals have two transverse red blotches, which almost merge in the center to become a band at mid length; the sepals are suffused with red at the tips, though sometimes this appears as a second blotch. The epichil of the lip is minutely spotted red all over with spots arranged along the veins. The fimbriae of the crest are dark red. The column ventral surface has minute red spots, the column is otherwise white or yellow, with green apical margins. The sepals and petals are elliptic and acuminate. The lip epichil is ovate and apiculate. The column wings are large, earlike, longer than wide, and angle outward. PLATE 48.

MEASUREMENTS Dorsal sepal 3.1 cm long, 1.8 cm wide; lateral sepals 3.5 cm long, 2 cm wide; petals 3.2 cm long, 2 cm wide; lip 2.3 cm long, 1.9 cm wide.

ETYMOLOGY Named for Gustav Wallis, who collected the type specimen.

DISTRIBUTION AND HABITAT Known from Colombia and Ecuador, at 650–1800 m in elevation.

FLOWERING TIME October to December.

Huntleya lucida (Rolfe) Rolfe, *Orch. Rev.* 24: 236. 1916. Basionym: *Zygopetalum lucidum* Rolfe, *Gard. Chron.*, ser. 3, 5(1): 799. 1889. Type: Guyana (formerly British Guiana) (holotype: K).

DESCRIPTION The sepals and petals are basally white. At one-third their length they have two large red-brown blotches on each side of the midvein followed by a central band that is green, then red-brown again; the apexes of the sepals and petals are light yellow-green. The lip hypochil is white, and the epichil is dark red-brown grading to white at the apex. The callus is white and the fimbriae and callus have brown nerves. The column is white with dark purple longitudinal lines near its base. There is no quilting on the sepals and petals. The lip is simple or inconspicuously three-lobed; the basal lobes are very small and concave. The surface of the lip is glandular, like the texture of velvet. The callus crest is in the form of a semicircle, concave, with filamentous appendages that radiate from the callus crest; the most apical part of the crest has fimbriae that are much shorter than the lateral fimbriae or this area can be without any fimbriae. The column is winged and the wings do not angle outwardly. PLATE 49.

COMMENT The range of this species is broad. Dodson (1980) discussed the possibility that the Ecuadorian specimens bridge the differences between *Huntleya fasciata* and *H. lucida*, questioning the validity of *H. fasciata* as a separate species. There doesn't seem to be much difference in the flowers of the two species other than the dimensions of the lip, and perhaps the length of the apical-most fimbriae.

MEASUREMENTS Leaves 13–25 cm long, 2.5–6.0 cm wide; inflorescence 18 cm long; sepals and petals 2.6–3.8 cm long, 0.7–1.2 cm wide, petals slightly shorter than sepals; lip 1.8–2.3 cm long, 1.2 cm wide; column 1.4 cm long.

ETYMOLOGY Latin *lucidus*, shining, referring to the waxy shine of the flower.

DISTRIBUTION AND HABITAT Northern Brazil (Roraima), Ecuador, Guyana, Honduras, and Venezuela, most probably over northern South America and Central America, at 600–650 m in elevation.

FLOWERING TIME January and February.

Huntleya meleagris Lindley, *Bot. Reg.* 23: plate 1991. 1837. Type: Brazil, based on drawing by M. Descourtilz (type: K); type species of *Huntleya*.

SYNONYMS

Batemannia meleagris Reichenbach f., *Bonplandia*
3: 217. 1855.

Zygopetalum meleagris Bentham, *Journ. Linn. Soc.
Bot.* 18: 321. 1880.

DESCRIPTION The sepals and petals are red-brown with some yellow spots and, if transverse barring is present, it is not distinct. The basal third of the sepals and petals is white or pale yellow. Going up from the base, color changes from white to orange, becoming brown-red apically. The lip hypochil is white or yellow and the apical half of the epichil is red-brown or brown-purple. The texture of the entire flower is waxy, and the sepals and petals are heavily quilted. The basal callus has an erect semicircular or square fimbriate crest, the posterior margins of which are incumbent on the column foot. The base of the hypochil has a second set of fimbriae only on the posterior half of the hypochil with the fimbriae pointing toward the column foot. The apical lobe of the lip is obscurely three-lobed and articulated with the apexes of the callus plate; the apex of the lip is recurved. The column is stout, erect, semiterete below, dilated above, with a conspicuous dorsal keel and broad triangular lateral wings which are confluent at the column apex. PLATE 50; FIGURE 3.

COMMENT *Huntleya meleagris* has a square-shaped crest on the hypochil of the lip with fimbriae on three sides, the posterior (or basal portion) of which is free of fimbriae. The crest of *H. lucida*, in contrast, is more rounded with the apical-most part of the crest free of fimbriae. *Huntleya meleagris* has more prominent wings on the column and a hint of three lobes to the lip epichil, and the lip is wider than *H. lucida*.

MEASUREMENTS Leaves 15–30 cm long, 3.0–4.5 cm wide; inflorescence 10–15 cm tall; dorsal sepal 4–6 cm long, 2.0–2.5 cm wide; lateral sepals 4.5–6.0 cm long, 1.5–2.5 cm wide; petals 3.5–5.6 cm long, 2–3 cm wide; lip 2.5–3.5 cm long, 2–3 cm wide; column 1.5–2.0 cm long.

ETYMOLOGY Afrikaans *mealie*, an ear of corn, referring to the surface of the sepals and petals; also said to be Greek for Guinea fowl, and a reference to the resemblance of the flower to the plumage of the fowl.

DISTRIBUTION AND HABITAT Known from Venezuela and Brazil along the coast, at 700–1000 m in elevation, in highland forests growing in deep shade. Reports of the species from Costa Rica, Nicaragua, Panama, Colombia, Ecuador, Bolivia, and Peru are probably of *Huntleya burtii*.

FLOWERING TIME Not recorded.

Huntleya sessiliflora Bateman ex Lindley, *Bot. Reg.* 23: plate 1991. 1837. Type: Guyana, collected by Robert Schomburgk (holotype: K).

DESCRIPTION The sepals and petals are white at the base, becoming yellow at the apex with, at about one-third of the length from the base, two brown blotches on either side of the midvein, then a yellow band, and next a broad brown area to the apex, with just the apical tips becoming yellow again. The lip is white basally with a yellow or red apex. The column is creamy green. There is minimal quilting on the sepals and petals. The lip epichil is three-lobed, the hypochil has two sets of fimbriae—an upper square-shaped crest, and a lower crest that is posterior on the hypochil and points toward the column foot.

COMMENT *Huntleya sessiliflora* was proposed by John Lindley in 1837 based on a communication with James Bateman, which Bateman doesn't recall and later rejected. Robert Rolfe went through Lindley's herbarium and found a flower ticketed "interior of British Guiana, Mr. Schomburgk"; the specimen was not labeled *H. sessiliflora*, but penciled on the sheet is the comment "probably *H. meleagris*."

Fowlie (1967) made the argument that the species was invalid, but seven years later he (Fowlie 1974) decided the species was valid

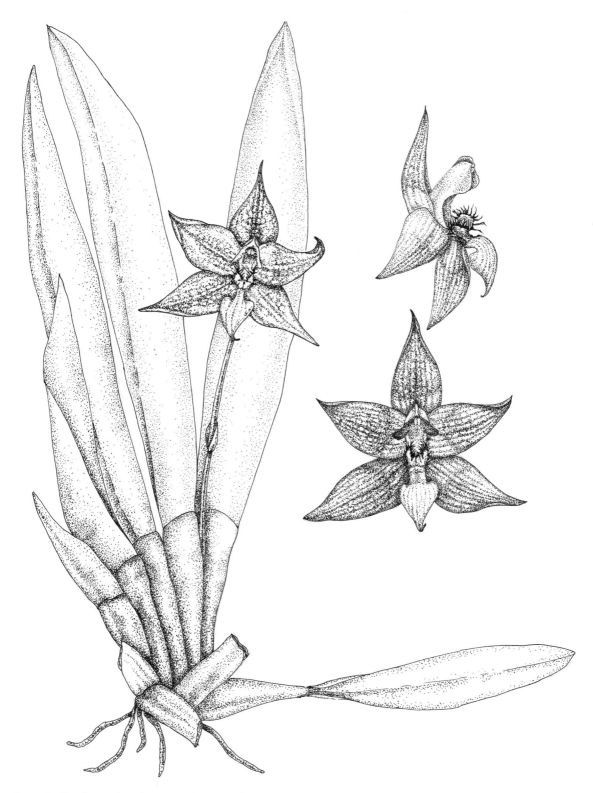

Figure 3. *Huntleya meleagris.* Drawing by Jane Herbst

because there was a holotype and it was published. In an article on the pages preceding Fowlie's 1974 article, Galfrid Dunsterville listed two species of *Huntleya* in Venezuela, *H. lucida* and *H. wallisii* (Dunsterville and Dunsterville 1974, 115–116). (In that article Dunsterville said that *H. meleagris* from Venezuela is more correctly *H. wallisii*. The *H. wallisii* he refers to has a flower 10 cm is diameter, unlikely for a true *H. wallisii*.) Fowlie (1974) believed the species was actually *H. sessiliflora*. He even took the step of relabeling Dunsterville's drawing of *H. meleagris* in *Venezuelan Orchids Illustrated* (Ill. 3: 146–147) as *H. sessiliflora* for his (Fowlie's) article. Fowlie presented photographs labeled *H. sessiliflora*. This does indeed look like the drawing of the *H. meleagris* in *Venezuelan Orchids Illustrated*. To me this photo fits in the parameters of *H. meleagris*, perhaps with thinner segments and paler colors.

The description and measurements given in the present volume are taken from the drawing of *Huntleya meleagris* in *Venezuelan Orchids Illustrated* by Dunsterville and Garay that was relabeled by Fowlie as *H. sessiliflora*. This then represents *H. sessiliflora sensu* Fowlie. Whether this is correct is unknown. I feel the truth probably is that *H. sessiliflora* should be included as a synonym of *H. meleagris*. Examination of the type of each would need to be done to determine this, and actually has been done by others with mixed results.

MEASUREMENTS Leaves 22 cm long, 5 cm wide; inflorescence 10 cm long; sepals and petals 5 cm long, 2 cm wide; lip 2.5 cm long, 1 cm wide; column 2 cm long.

ETYMOLOGY Latin *sessilis*, stalkless, and *florus*, flower, referring to the type specimen, which was a flower without a stem or peduncle.

DISTRIBUTION AND HABITAT Colombia, Ecuador, Guyana, Trinidad, and Venezuela, at 400–800 m in elevation.

FLOWERING TIME October to December.

Huntleya vargasii Dodson & D. E. Bennett, *Icon. Pl. Trop.*, ser. 2, 1: plate 80. 1989. Type: Peru, Junín, Chanchamayo, 1500 m, collected by A. Vargas, *Bennett 3885* (holotype: MO).

DESCRIPTION The apical half of each sepal is dark red-brown, the center is red-yellow with a lower border of yellow, and the basal quarter is white. The petals are similar in color but with two yellow lines (the original description text does not say where the lines are) and a brown blotch on the white petal base. The lip is white basally, the epichil is rose to purple-rose, and the callus base has pale rose-pink fimbriae. The column is white, the apex and margins are green, the column interior is streaked with purple, and the column base is dark purple. The anther is white with pale brown margins. There is minimal quilting on the surface of the sepals and petals. The lip is held erect and parallel to the column and at right angles to the sepals and petals. The epichil has 12 raised lamellae or possibly raised veins that begin at the base of the epichil and go to the apex of the lip.

COMMENT *Huntleya vargasii* may be a form of *H. burtii* that holds its lip horizontally rather than in the plane of the sepals and petals. Many people have sent me pictures of other *Huntleya* species showing a flower with the lip held horizontally. Whether this is a trait of an immature flower or a chance occurrence has been speculated. The veins on the lip, drawn so prominently on the type drawing, are also seen on *H. burtii* and may just be an artistic interpretation that is not truly something that is more prominent on *H. vargasii*. These arguments would favor *H. vargasii* being a synonym of *H. burtii*, or an intermediate of *H. burtii* and *H. wallisii*.

MEASUREMENTS Leaves 20–33 cm long, 4.5–7.5 cm wide; inflorescence 16 cm long; dorsal sepal 4.4 cm long, 2.1 cm wide; lateral sepals 5 cm long, 2 cm wide; petals 3.8 cm long, 2.4 cm wide; lip 2.6 cm long, 1.5 cm wide; column 2.7 cm long, 1.3 cm wide.

ETYMOLOGY Named for Antonieta Vargus P., who collected and grew the type plant.

DISTRIBUTION AND HABITAT Known only from Peru, at 1500 m in elevation, in tropical wet forest.

FLOWERING TIME October to June.

Huntleya wallisii (Reichenbach f.) Rolfe, *Orch. Rev.* 24: 236. 1916. Basionym: *Batemannia burtii* var. *wallisii* Reichenbach f., *Gard. Chron.* 575. 1873. Type: Colombia, *G. Wallis s.n.* (holotype: W).

SYNONYMS
Batemannia wallisii (Reichenbach f.)
   Reichenbach f., *Gard. Chron.* 1: 776. 1880.
Huntleya burtii var. *wallisii* (Reichenbach f.)
   Rolfe, *Orch. Rev.* 8: 272. 1900.
Zygopetalum burtii var. *wallisii* (Reichenbach f.)
   Veitch, *Man. Orch. Pl.* 2: 45. 1887.

DESCRIPTION The sepals are chestnut brown to red with white at the base, while the petals are chestnut brown to red with yellow at the base and have a brown-purple blotch at the base that is broken into lines. The lip is a darker brown to red than the other segments, and the crest is a white plate fringed with long narrow red-brown teeth. The second set of fimbriae at the base of the hypochil is also red, and a red line separates the hypochil from the epichil. The column is pale green with a purple spot on each side of the stigma. There is minimal or little quilting on the sepals and petals, which are narrow for the genus. The lip is long and narrow. The crest is fimbriate with all the fimbriae pointed upward, and there is a second line of fimbriae that comes out around the bottom of the callus plate, creating a downward-facing skirt. On some specimens this skirt is only posterior and does not wrap around the junction of the hypochil and epichil. The column wings are rounded and angle down instead of out. PLATE 51.

COMMENT *Huntley wallisii* is very similar to *H. burtii*, but has much longer, relatively thinner sepals and petals. The lip is much narrower and longer than that of *H. burtii*, and the column wings are rounded and down-turned, compared to the column wings of *H. burtii*, which are pointed and upswept. *Huntleya wallisii* has a much bigger and showier flower with somewhat different patterns of color. It is the only *Huntleya* species with a red line at the junction of the hypochil and epichil, and this line has fine barely visible hairs. *Huntley wallisii* too may just be a form or variety of *H. burtii*, as there are many intermediate specimens.

MEASUREMENTS Leaves 30 cm long, 2 cm wide; inflorescence 16 cm long; sepals 7 cm long, 2.2 cm wide; petals 6.8 cm long, 3 cm wide; lip 3.5 cm long, 2 cm wide; column 2.5 cm long.

ETYMOLOGY Named for Gustav Wallis, who collected the type specimen.

DISTRIBUTION AND HABITAT Known from Colombia and Ecuador, at 300–800 m in elevation, in tropical wet forest.

FLOWERING TIME Throughout the year.

Ixyophora Dressler, *Lankesteriana* 5(2): 95. 2005.

Ixyophora species were initially included in the genus *Chondrorhyncha*, using its older definitions. Now with the formation of *Chondrorhyncha sensu stricto*, this group of species required a new name. Robert Dressler created *Ixyophora* in response to molecular data as well as morphological similarities.

The plants of *Ixyophora* lack pseudobulbs. The dark green leaves are oblanceolate (widest at three-quarters length) and taper to the conduplicate petiole base. The resupinate flowers on pendent inflorescences are yellow or green with various yet minimal markings. The dorsal sepal is either concave or convex. The lateral sepals reflex backward to varying degrees and curl along their long margins, forming false nectaries and in some species appearing hooklike at their apexes. The lip, though tubular, does not cover the column. The callus is platelike, covering the base of the lip and extending laterally to the lateral walls of the lip. There is a slight sulcus below the apical end of the callus. The callus has keels that, to varying degrees, run from the lip base to the apical margin of the callus. The clublike column widens significantly at the slitlike stigma, and the clinandrium is on the ventral side of the column. Some species have a ventral tooth on the long column foot. The form of the stipe is short and narrow between the viscidium and pollinaria, and the viscidium is trullate.

Ixyophora is sister to the genera *Aetheorhyncha* and *Chaubardiella*. At first glance, the yellow tubular flowers with reflexed sepals are reminiscent of the Mexican species so common in collections in the United States, *Stenotyla lendyana*, but the lip of *Stenotyla* clasps the column and the callus is narrow, among other differences. The molecular data reveal that *Ixyophora* and *Stenotyla* are spaced apart from each other in the cladogram, with other genera between them.

Etymology
Greek *ixys*, waist, and *phoreus*, bearer or carrier, referring to the narrow waist of the stipe.

List of the Species of *Ixyophora*

Ixyophora aurantiaca

*Ixyophora carinata**

Ixyophora fosterae

Ixyophora luerorum

*Ixyophora viridisepala** (type species of *Ixyophora*)

* Molecular sampling confirms placement of this species within the genus
  Ixyophora.

Key to the Species of *Ixyophora*

1a. Callus midrib extends from base of lip to the apex of callus, flowers green
with white, lip with some purple pigment on callus . . . . *I. viridisepala*

1b. Callus midrib does not extend from base of lip to apex, flowers yellow,
callus without any purple pigment . . . . go to 2

2a. Lip four-lobed (two lateral lobes obscure), yellow without spots . . . .
. . . .*I. carinata*

2b. Lip one-lobed, yellow with spots . . . . go to 3

3a. Callus extends to three-quarter lip length . . . . *I. fosterae*

3a. Callus extends to half of lip length or less. . . . . go to 4

4a. Callus extends to midquarter lip length . . . .*I. aurantiaca*

4b. Callus extends to half of lip length. . . . .*I. luerorum*

Ixyophora aurantiaca (Senghas & G.
Gerlach) Dressler, *Lankesteriana* 5(2): 95. 2005.
Basionym: *Chondrorhyncha aurantiaca* Senghas
& G. Gerlach, *Orchidee (Hamburg)* 42(6): 282.
1991. Type: Peru, collected by B. Würstle &
M. Arrias-Silva, *Botanic Garden Heidelberg O-
19869* (holotype: HEID).

DESCRIPTION The sepals and petals are green-
yellow to yellow, the lip is dark yellow with red-
brown to orange spots arranged along the veins
of the lip and callus, the lip margins are free
of spots. The lateral sepals are mildly inrolled
along their length and reflexed backward
slightly. The petals and dorsal sepal loosely
cover the column. The one-lobed lip has lateral
edges which curl up to but do not cover the column. The apical half of the lip margin is erose
to slightly crispate. The callus is slightly raised
apically, with two toothlike projections apically. PLATE 52.

COMMENT *Ixyophora aurantiaca* may be a synonym or variety of *I. luerorum*. The major difference between the two species is the longer callus of *I. luerorum*.

MEASUREMENTS Leaves 12 cm long, 3 cm wide;
inflorescence 6–7 cm long; dorsal sepal 2 cm
long, 0.9 cm wide; lateral sepals 2.5 cm long,
0.9 cm wide; petals 2 cm long, 1.2 cm wide; lip 2
cm long and wide; column 1.2 cm long.

ETYMOLOGY Latin *aurantiacus*, between yellow
and scarlet, referring to the flower color.

DISTRIBUTION AND HABITAT Known only from
Peru.

FLOWERING TIME October to December.

Ixyophora carinata (P. Ortíz) Dressler,
Lankesteriana 5(2): 95. 2005. Basionym:
Chondrorhyncha carinata P. Ortíz, *Orquideología*
19: 18. 1994. Type: Colombia, Amazonas,
collected by L. A. Serna, July 1991, flowered
in cultivation at Colomborquídeas, *R. Escobar
5433* (holotype: UJCOL).

DESCRIPTION The flowers are yellow. The dorsal sepal and petals hood over the column. The lateral sepals reflex backward. The lip is four-lobed with small apical lobes. The apical lip margins are undulate. The callus is variably two- or three-toothed or irregularly toothed with curved lateral margins. Veins run from the lip base to the margins of the callus, but the mid keel starts at mid callus length. The column has a minutely papillose keel or ridge on the ventral side of the column. The rostellum is five-toothed with the central tooth longest. PLATE 53.

COMMENT Plants with tubular flowers that are yellow could belong to several species in various genera of the *Huntleya* alliance, such as *Benzingia thienii*, *Chondrorhyncha fosterae*, *Daiotyla xanthina*, *Ixyophora carinata*, and *Stenotyla lendyana*. Careful attention to the callus will determine which genus is correct.

MEASUREMENTS Leaves 15 cm long, 3 cm wide; inflorescence 4–6 cm long; dorsal sepal 1.6 cm long, 0.5 cm wide; lateral sepals and petals 2 cm long, 0.9 cm wide; lip 2.2 cm long and 1.5 wide; column 1.2 cm long

ETYMOLOGY Latin *carinatus*, keeled, describing the ridge on the ventral side of the column.

DISTRIBUTION AND HABITAT Known from Colombia, elevation not recorded.

FLOWERING TIME February to July.

Ixyophora fosterae (Dodson) P. A. Harding, *comb. nov.* Basionym: *Chondrorhyncha fosterae* Dodson, *Selbyana* 7(2–4): 357. 1984, as "*Chrondrorhyncha*." Type: Bolivia, Cochabamba, San Onofre, 1800 m, flowered in cultivation 27 September 1982, *M. Foster s.n.* (holotype: SEL).

DESCRIPTION The sepals and petals are yellow-green, and the lip is yellow with red spots at the base. The lateral sepals reflex back. The lip has a denticulate apical margin. The callus is a raised plate that extends transversely almost

to the lateral edges of the lip margin with five to seven lamellae that go to or are apical to the mid length of the lip; the central lamella is raised. The column is widest at the stigma with lateral wings and the column ventral surface is warty or hairy. PLATE 54.

COMMENT Calaway Dodson labeled the plant in plate 517 of *Icones Plantarum Tropicarum* as *Chondrorhyncha rosea* Lindley, but corrected it to *C. fosterae* in *Selbyana* (1984, 7[2–4]: 357).

MEASUREMENTS Leaves 15–20 cm long, 2.5 cm wide; inflorescence 12 cm long, with one or two bracts; dorsal sepal 1.5 cm long, 0.6 cm wide; lateral sepals 2.0–2.5 cm long, 0.8 cm wide; petals 2.3 cm long, 1.2 cm wide; lip 2.3 cm long, 2 cm wide; column 1 cm long.

ETYMOLOGY Named to honor Dr. Mercedes Foster, who collected the type species.

DISTRIBUTION AND HABITAT Known only from Bolivia, at 1300–1800 m in elevation.

FLOWERING TIME Recorded in March and September.

Ixyophora luerorum (R. Vásquez & Dodson) P. A. Harding, *comb. nov.* Basionym: *Chondrorhyncha luerorum* R. Vásquez & Dodson, *Rev. Soc. Boliv. Bot.* 2(1): 1. 1998. Type: Bolivia, Cochabamba, Carrasca, Sehuencas, 2250 m, 14 January 1988, *R. Vásquez, C. Luer & J. Luer 1003* (holotype: LPB; isotype: Herb. Vasquezianum).

DESCRIPTION The sepals and petals are green-yellow, the lip is dark yellow with red spots diffusely spread over the callus and overlaid red apical to the callus, sparing the lip apexes. The lateral sepals are held laterally. The one-lobed lip has lateral edges which curl up to but do not cover the column. The apical half of the lip margin is erose to slightly crispate. The callus extends apically to nearly half the length of the lip and is slightly raised apically, with teethlike projections apically. PLATE 55.

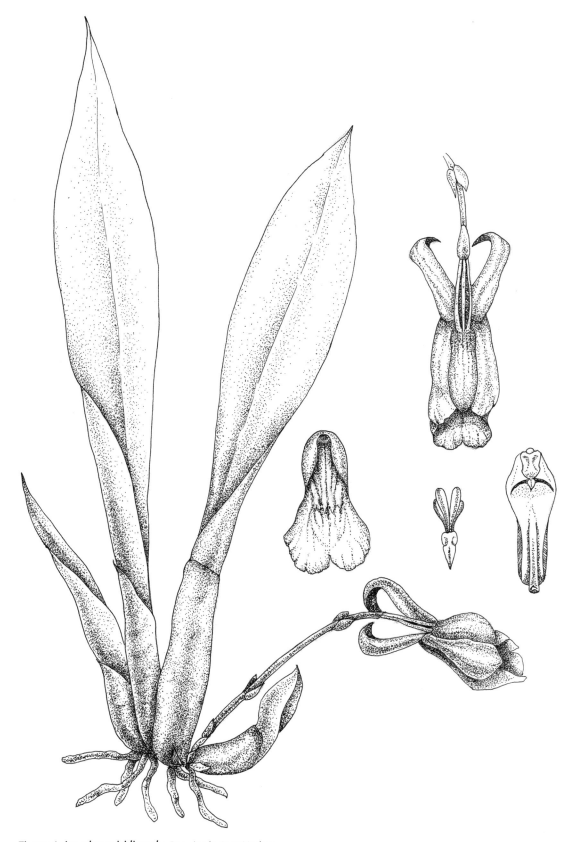

Figure 4. *Ixyophora viridisepala.* Drawing by Jane Herbst

COMMENT *Ixyophora aurantiaca* may be a synonym or variety of *I. luerorum*. The major differences between the two are the longer callus of *I. luerorum* that extends to nearly the mid length of the lip, and the red color of *I. luerorum* being more diffuse over the lip.

MEASUREMENTS Leaves 18–26 cm long, 2–4 cm wide; inflorescence 10 cm long; dorsal sepal 2.5 cm long, 1.1 cm wide; lateral sepals 3 cm long, 2 cm wide; petals 2.5 cm long, 1.5 cm wide; lip 2.5 cm long, 2.4 cm wide; column 1.5 cm long.

ETYMOLOGY Named for Carl and Jane Luer, who helped in the description of the species.

DISTRIBUTION AND HABITAT Known from Bolivia, at 2150–2350 m, in cloud forests.

FLOWERING TIME December and January.

Ixyophora viridisepala (Senghas) Dressler,
Lankesteriana 5(2): 95. 2005. Basionym: *Chondrorhyncha viridisepala* Senghas, *Orchidee (Hamburg)* 40: 181. 1989. Type: Ecuador, Morona-Santiago, 1300 m, collected by K. Senghas, L. Bockemühl & H. Volles, *Botanical Garden Heidelberg sub 0-19022* (holotype: HEID); type species of *Ixyophora*.

DESCRIPTION The sepals and petals are clear yellow-green with sepals that are slightly darker green. The lip is pale green in the middle, the lip base is green-yellow, the lip apex is pale green-white, sometimes with diffuse minute red spots. The callus midrib is purple. The column is green-white; the base and foot of the column are pale green with purple dots on the ventral surface and with a small purple spot on each side of the stigma. The anther is white and lightly green in the center and the pollinia are pale yellow. The dorsal sepal is concave. The lateral sepals are retuse, sweeping back and hook-like; the interior of the sepals is pubescent at the base. The petals are rounded, very shallowly emarginate. The lip is unlobed and tubular; the lip base is subsaccate while the apical lip margin is emarginate and lightly undulate. The callus has five raised keels with only the midrib extending fully from the lip base to the transverse ridge in the middle of the lip; the other ribs begin mid length from the lip base to the ridge. The column is straight, clavate, obscurely winged, and papillose at the base. PLATE 56; FIGURE 4.

MEASUREMENTS Leaves 12–18 cm long, 2.3 cm wide; inflorescence 8 cm long; dorsal sepal 2.1 cm long, 0.8 cm wide; lateral sepals 2.6 cm long, 0.8 cm wide; petals 2.7 cm long, 1.2 cm wide; lip 3.4 cm long, 2.1 cm wide; column 1.2 cm long, 0.7 cm wide.

ETYMOLOGY Latin *viridi*, green, and *sepalus*, sepaled, referring to the color of the sepals.

DISTRIBUTION AND HABITAT Known only from Ecuador and Peru, at 800–1950 m in elevation, in wet montane forest.

FLOWERING TIME April and May.

Plate 1. *Aetheorhyncha andreettae.* Photo by Manfred Speckmaier

Plate 2. *Benzingia caudata.* Photo by Patricia Harding

Plate 3. *Benzingia estradae.* Photo by Manfred Speckmaier

Plate 4. *Benzingia hajekii.* Photo by Manfred Speckmaier

Plate 5. *Benzingia hirtzii.* Photo by Eric Hunt

Plate 6. *Benzingia jarae.* Photo by Günther Gerlach

Plate 7. *Benzingia palorae.* Photo by Günther Gerlach

Plate 8. *Benzingia reichenbachiana.* Photo by Patricia Harding

Plate 9. *Chaubardia heteroclita.* Photo by Patricia Harding

Plate 10. *Chaubardia klugii.* Photo by Ron Parsons

Plate 11. *Chaubardiella chasmatochila.* Photo by Patricia Harding

Plate 12. *Chaubardiella dalessandroi.* Photo by Ron Parsons

Plate 13. *Chaubardiella hirtzii*. Photo by Günther Gerlach

Plate 14. *Chaubardiella pacuarensis*. Photo by Manfred Speckmaier

Plate 15. *Chaubardiella pubescens*. Photo by Ron Parsons

Plate 16. *Chaubardiella subquadrata*. See text for opinions on this photograph.
Photo by Ron Parsons

Plate 17. *Chaubardiella tigrina.* Photo by Ron Parsons

Plate 19. *Chondrorhyncha manzurii.* Photo by Francisco Villegas

Plate 18. *Chondrorhyncha hirtzii.* Photo by Ron Parsons

Plate 20. *Chondrorhyncha panguensis.* Photo by Alex Portillo

Plate 22. *Chondrorhyncha suarezii.* Photo by Alex Hirtz

Plate 21. *Chondrorhyncha rosea.* Photo by Günther Gerlach

Plate 23. *Chondrorhyncha velastigui.* Photo by Alex Portillo

Plate 24. *Chondroscaphe amabilis.* Photo by Patricia Harding

Plate 25. *Chondroscaphe atrilinguis.* Photo by Gary Yong Gee

Plate 26. *Chondroscaphe chestertonii.* Photo by Patricia Harding

Plate 27. *Chondroscaphe dabeibaensis.* Photo by Patricia Harding

Plate 28. *Chondroscaphe eburnea.* Photo by Günther Gerlach

Plate 29. *Chondroscaphe embreei.* Photo by Patricia Harding

Plate 30. *Chondroscaphe endresii.* Photo by Manfred Speckmaier

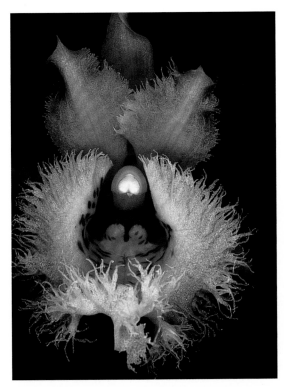

Plate 31. *Chondroscaphe escobariana.* Photo by Ron Parsons

Plate 32. *Chondroscaphe flaveola.* Photo by Günther Gerlach

Plate 33. *Chondroscaphe gentryi.* Photo by Ron Parsons

Plate 34. *Chondroscaphe laevis.* Photo by Rudolf Jenny

Plate 35. *Chondroscaphe merana.* Photo by Günther Gerlach

Plate 36. *Cochleanthes flabelliformis.* Photo by Manfred Speckmaier

Plate 37. *Daiotyla xanthina.* Photo by Patricia Harding

Plate 39. *Echinorhyncha ecuadorensis.* Photo by Alex Hirtz

Plate 38. *Echinorhyncha antonii.* Photo by Carlos Uribe Vélez

Plate 41. *Echinorhyncha vollesii.* Photo by Alex Hirtz

Plate 40. *Echinorhyncha litensis.* Photo by Manfred Speckmaier

Plate 42. *Euryblema anatonum.* Photo by Manfred Speckmaier

Plate 43. *Euryblema andreae.* Photo by Manfred Speckmaier

Plate 44. *Hoehneella gehrtiana*. Photo by Günther Gerlach

Plate 45. *Huntleya burtii*. Photo by Patricia Harding

Plate 46. *Huntleya caroli*. Photo by Carlos Uribe Vélez

Plate 47. *Huntleya citrina*. Photo by Ron Parsons

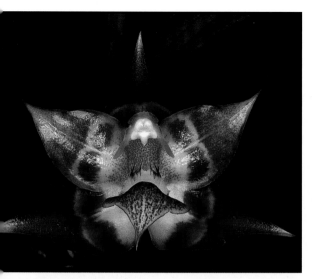

Plate 48. *Huntleya gustavi*. Photo by Ron Parsons

Plate 49. *Huntleya lucida*. Photo by Günther Gerlach

Plate 50. *Huntleya meleagris*. Photo by Patricia Harding

Plate 51. *Huntleya wallisii*. Photo by Patricia Harding

Plate 52. *Ixyophora aurantiaca*. Photo by Günther Gerlach

Plate 53. *Ixyophora carinata*. Photo by Patricia Harding

Plate 54. *Ixyophora fosterae*. Photo by Manfred Speckmaier

Plate 55. *Ixyophora luerorum*. Photo by Günther Gerlach

Plate 56. *Ixyophora viridisepala*. Photo by Ron Parsons

Plate 57. *Kefersteinia alata*. Photo by Ron Parsons

Plate 58. *Kefersteinia alba*. Photo by Eric Hunt

Plate 59. *Kefersteinia andreettae*. Photo by Patricia Harding

Plate 60. *Kefersteinia auriculata.* Photo by Gary Yong Gee

Plate 61. *Kefersteinia aurorae.* Photo by Patricia Harding

Plate 62. *Kefersteinia bertoldii.* Photo by Rudolf Jenny

Plate 63. *Kefersteinia candida.* Photo by Alex Hirtz

Plate 64. *Kefersteinia chocoensis.* Photo by Günther Gerlach

Plate 65. *Kefersteinia costaricensis.* Photo by Ron Parsons

Plate 66. *Kefersteinia elegans.* Photo by Ron Parsons

118

Plate 67. *Kefersteinia escalerensis.* Photo by Patricia Harding

Plate 68. *Kefersteinia escobariana.* Photo by Tilman Neudecker

Plate 69. *Kefersteinia excentrica.* Photo by Ron Parsons

Plate 70. *Kefersteinia expansa*. Photo by Patricia Harding

Plate 71. *Kefersteinia forcipata*. Photo by Alex Portillo

Plate 72. *Kefersteinia gemma*. Photo by Marni Turkel

Plate 73. *Kefersteinia graminea*. Photo by Günther Gerlach

Plate 74. *Kefersteinia guacamayoana.* Photo by Alex Hirtz

Plate 75. *Kefersteinia hirtzii.* Photo by Alex Hirtz

Plate 76. *Kefersteinia jarae.* Photo by Carlos Uribe Vélez

Plate 77. *Kefersteinia koechliniorum.* Photo by Ron Parsons

Plate 78. *Kefersteinia lactea*. Photo by Ron Parsons

Plate 79. *Kefersteinia lafontainei*. Photo by Günther Gerlach

Plate 80. *Kefersteinia laminata*. Photo by Patricia Harding

Plate 81. *Kefersteinia licethyae*. Photo by Alex Hirtz

Plate 82. *Kefersteinia lindneri.* Photo by Ron Parsons

Plate 83. *Kefersteinia microcharis.* Photo by David Hunt

Plate 84. *Kefersteinia mystacina.* Photo by Ron Parsons

Plate 85. *Kefersteinia niesseniae.* Photo by Ron Parsons

Plate 86. *Kefersteinia ocellata*. Photo by Ron Parsons

Plate 87. *Kefersteinia orbicularis.* Photo by Ron Parsons

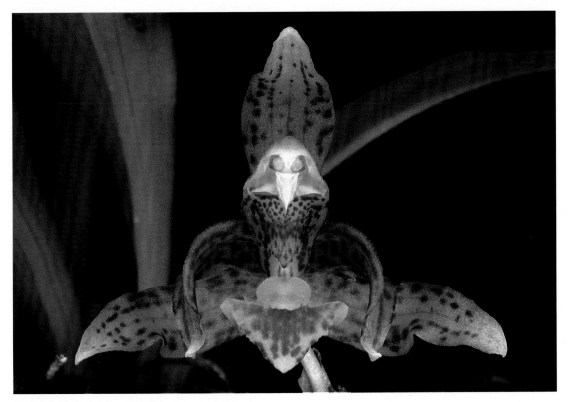

Plate 88. *Kefersteinia parvilabris.* Photo by Ron Parsons

126

Plate 89. *Kefersteinia perlonga*. Photo by Ron Parsons

Plate 90. *Kefersteinia pseudopellita*. Photo by Patricia Harding

Plate 91. *Kefersteinia pulchella*. Photo by Günther Gerlach

Plate 92. *Kefersteinia pusilla*. Photo by Carlos Hajek

Plate 93. *Kefersteinia retanae.* Photo by Manfred Speckmaier

Plate 94. *Kefersteinia sanguinolenta.* Photo by Günther Gerlach

Plate 95. *Kefersteinia stevensonii.* Photo by Ron Parsons

Plate 97. *Kefersteinia taurina.* Photo by Rudolf Jenny

Plate 96. *Kefersteinia taggesellii.* Photo by Ron Parsons

128

Plate 98. *Kefersteinia tinschertiana*. Photo by Moises Béhar

Plate 99. *Kefersteinia tolimensis*. Photo by Patricia Harding

Plate 100. *Kefersteinia trullata*. Photo by Ron Parsons

Plate 101. *Kefersteinia urabaensis*. Photo by Patricia Harding

Plate 102. *Kefersteinia villenae.* Photo by Carlos Hajek

Plate 103. *Kefersteinia vollesii.* Photo by Rudolf Jenny

Plate 104. *Kefersteinia wercklei.* Photo by Günther Gerlach

Plate 105. *Pescatorea cerina*. Photo by Ron Parsons

Plate 106. *Pescatorea cochlearis*. Photo by Patricia Harding

Plate 107. *Pescatorea coelestis*. Photo by Patricia Harding

Plate 108. *Pescatorea coronaria*. Photo by Patricia Harding

Plate 109. *Pescatorea dayana*. Photo by Patricia Harding

Plate 110. *Pescatorea ecuadorana*. Photo by Alex Portillo

Plate 111. *Pescatorea hemixantha*. Photo by Manfred Speckmaier

Plate 112. *Pescatorea hirtzii*. Photo by Ron Parsons

Plate 113. *Pescatorea klabochorum.* Photo by Patricia Harding

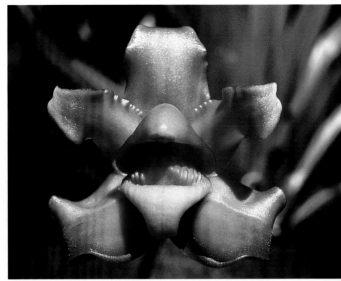

Plate 114. *Pescatorea lalindei.* Photo by Ron Parsons

Plate 115. *Pescatorea lamellosa.* Photo by Patricia Harding

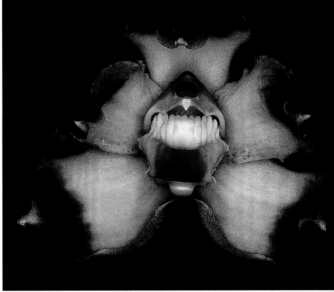

Plate 116. *Pescatorea lawrenceana,* **open form.** Photo by Ron Parsons

Plate 117. *Pescatorea lehmannii*. Photo by Günther Gerlach

Plate 118. *Pescatorea pulvinaris*. Photo by Ron Parsons

Plate 119. *Pescatorea violacea*. Photo by Patricia Harding

Plate 120. *Pescatorea wallisii.* Photo by Manfred Speckmaier

Plate 121. *Stenia angustilabia.* Photo by Manfred
Speckmaier

Plate 122. *Stenia aurorae.* Photo by Ron Parsons

136

Plate 123. *Stenia bismarckii.* Photo by Ron Parsons

Plate 125. *Stenia calceolaris.* Photo by Ron Parsons

Plate 124. *Stenia bohnkiana.* Photo by Vitorino Paiva Castro Neto

Plate 126. *Stenia guttata.* Photo by Ron Parsons

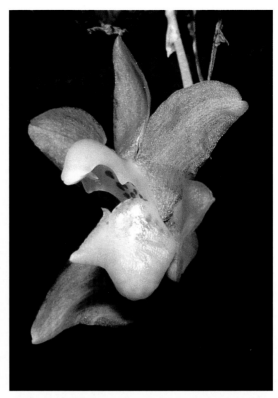

Plate 127. *Stenia jarae.* Photo by Manfred Speckmaier

Plate 128. *Stenia lillianae.* Photo by Ron Parsons

Plate 129. *Stenia luerorum.* Photo by Manfred Speckmaier

Plate 130. *Stenia pallida.* Photo by Ron Parsons

Plate 131. *Stenia pastorellii.* Photo by Ron Parsons

Plate 132. *Stenia pustulosa.* Photo by Manfred Speckmaier

Plate 133. *Stenia saccata*. Photo by Alex Portillo

Plate 134. *Stenia stenioides*. Photo by Eric Christenson

Plate 135. *Stenia uribei*. Photo by Carlos Uribe Vélez

Plate 136. *Stenia vasquezii*. Photo by Ron Parsons

Plate 137. *Stenotyla helleri*. Photo by Rudolf Jenny

Plate 139. *Stenotyla lendyana*. Photo by Ron Parsons

Plate 138. *Stenotyla lankesteriana*. Photo by Ron Parsons

Plate 140. *Stenotyla picta*. Photo by Günther Gerlach

Plate 141. *Warczewiczella amazonica*. Photo by Ron Parsons

Plate 142. *Warczewiczella candida*. Photo by Patricia Harding

Plate 143. *Warczewiczella guianensis.* Photo by Günther Gerlach

Plate 144. *Warczewiczella ionoleuca.* Photo by Ron Parsons

Plate 145. *Warczewiczella lipscombiae.* Photo by Günther Gerlach

Plate 146. *Warczewiczella lobata.* Photo by Patricia Harding

Plate 147. *Warczewiczella marginata.* Photo by Ron Parsons

Plate 148. *Warczewiczella palatina.* Photo by Günther Gerlach

Plate 149. *Warczewiczella timbiensis.* Photo by Gustavo Aguirre

Plate 150. *Warczewiczella wailesiana.* Photo by Ron Parsons

Kefersteinia Reichenbach f., *Bot. Zeit. (Berlin)* 10: 633. 1852.

Since its original description by Heinrich Gustav Reichenbach, the circumscription of the genus *Kefersteinia* has remained fairly stable. Molecular analysis of six species confirms it to be monophyletic and a close relative of *Benzingia, Daiotyla, Euryblema,* and *Stenia,* and a little more distant relative to *Echinorhyncha.*

Kefersteinia is such a large genus that it would be nice if it could be broken into more taxonomically user-friendly groups. Probably the most useful segregation of *Kefersteinia* species into groupings was proposed by Karl Senghas and Günther Gerlach in 1992. They created two sections of *Kefersteinia* based on the shape of the callus. *Kefersteinia* section *Papilionatae* Senghas & G. Gerlach, with *K. graminea* as the type species, encompassed 26 species. These all have a low flat callus of two lobes with two apical points and a median depression between the two lobes, like a butterfly with its wings slightly flexed. Senghas and Gerlach included in this first group *K. alba, K. andreettae, K. bertoldii, K. bismarckii, K. elegans, K. expansa, K. gemma, K. hirtzii, K. lactea, K. lafontainei, K. laminata, K. lindneri, K. lojae, K. minutiflora, K. mystacina, K. pastorellii, K. pellita, K. pulchella, K. pusilla, K. sanguinolenta, K. stapelioides, K. stevensonii, K. taurina,* and *K. tolimensis.* The second section, *Kefersteinia* section *Umbonatae* Senghas & G. Gerlach, with the type species *K. wercklei,* included nine species with a fleshy solid callus on a stalk. In this group are *K. auriculata, K. chocoensis, K. costaricensis, K. guacamayoana, K. maculosa, K. ocellata, K. parvilabris,* and *K. vollesii.*

Darius Szlachetko (2003: 335) elevated section *Umbonatae* to genus level as *Senghasia,* with *Kefersteinia wercklei* as the type species. He added *K. perlonga, K. retanae,* and *K. trullata* to the new genus. Szlachetko and Romowicz (2006) added *K. bertoldii, K. jarae, K. lactea,* and *K. licethyae* to the list of species included in *Senghasia.* (Inclusion of *K. lactea* species is perhaps a mistake by the authors. I suspect they meant to include *K. alata* instead.)

Molecular data do not support dividing the group into sections or different genera but dividing *Kefersteinia* does help make identification of the individual species easier. I have arranged the key below taking into consideration the two callus types.

Plants of *Kefersteinia* are epiphytic. The fan-shaped growths are close together along the rhizome, resulting in a clump of multiple growths. There are no pseudobulbs. The linear leaves are the thinnest, have the least substance, and are the most flexible of this *Huntleya* clade. The single flowers on semipendent inflorescences are medium to small compared to other flowers of the *Huntleya* clade. The small flower size is well compensated for by the tendency of most *Kefersteinia* species to send out simultaneously multiple inflorescences from the base of the plants. Many species keep sending out more inflorescences for many months. In cultivation, this results in a clump of plant growths ringed in flowers. Sometimes there will be two flowers per inflorescence but usually the second bud aborts. The sepals and petals are often similar and generally spreading, the lateral sepals are basally adnate to the column foot, forming a short mentum, and the petals are inserted obliquely into the column.

The flower typically has a lip that is flexibly attached to a column foot with or without a small isthmus often called the claw of the lip. The variably shaped lips fall into three groups. One group has simple lips that are pinched or folded transversely at the lip mid length; this group includes the species shaped similar to *Kefersteinia pusilla*. The second group has showy lips that fold forward or deflex either down or back sharply at mid length, forming a showy skirt; this group includes *K. gemma*, *K. pulchella*, and *K. tolimensis*. (See lists at end of this chapter.) The third group of species has concave lips that are very small in relation to size of the petals; it includes *K. ocellata*, *K. parvilabris*, and *K. stevensonii*. This last group of species has flowers that appear to be insect or fungal mimics. I have tried to keep the various groups together in the key, though I am not sure this means they are related closely.

Kefersteinia species always have a prominent callus near (sub-basal) or at the base of the lip (basal), and the shape of the callus is important in identification. Dressler (1993b) says the callus produces a perfume that is gathered by male euglossine bees. Whether the callus is the attractant or the callus and column direct the pollinator, the callus of *Kefersteinia* is clearly an important part of what makes one species different from another. The callus is typically bilobed but otherwise either stalked like a mushroom with lobules atop the stalk or formed by lamellae of various shapes, with or without points, with apical portions that are free of the floor of the lip disc to varying degrees.

The column has a basal foot that attaches to the lip; generally the foot is at a lax right angle to the column. The column is semiconic in cross section, generally with a narrow base, and has a swelling somewhere along the lateral sides below the stigma; sometimes the swelling forms projections or wings.

The stigma is slitlike and often associated with an infrastigmatic median ventral keel that in some species is shaped like a tooth on the underside (ventral side) of the column. This ventral keel or ridge is critical in pollination, as it forces the bee to go to one side of the flower when it backs out, which places the pollinaria on the bee's antenna base (Dressler 1993b). The column and the callus in almost all species are adpressed so that the column seems to rest on the callus, a feature that must be important in directing the pollinator's path. The four pollinia are unequal in size, compressed in two pairs on a large common viscidium. The viscidium is transversely deflexed at the middle forming a skirt.

Kefersteinia has the most species of the *Huntleya* clade. Many of them are frequently in cultivation and well known at shows and exhibits; however, the genus is also full of mislabeled plants and mislabeled photographs. It has been a challenge to sort this genus out. No one is familiar with all of the *Kefersteinia* species, and I have tried to pool as many resources and opinions as I could. It was tempting to combine many of the species as synonyms of other species. I used the rule that if there was a visible physical difference between species, they were left as separate species.

As specimens are collected, species are described one at a time with the information one has at hand at the time. It is better to have at least recorded the specimen and its differences than to do nothing and let the information be lost. But this situation has resulted in several species in this genus that probably should have been considered varieties or forms. Someone who has seen lots of the various species could potentially sort this out. I am not sure such a person exists today. I have tried to include other authors' combinations in the text.

Identification of the species in this group is made easier by knowing several things: the country of origin, the shape of the callus, and the shape of the column, especially if there are wings on the column, where the wings are along the column lateral edge, and if there is a ventral keel on the ventral surface of the column. So many *Kefersteinia* species can only be told apart by bending the lip back to reveal these features (yes, I am the one who has been bending back flower lips to photograph these aspects of the flowers—occasionally the lip breaks—excuse me). The key following helps sort through the list of species if you need to identify a specimen, but at the end of the chapter is a list of the species with showy skirts in groups based on their usual color patterns. The trouble with using color patterns for identification is that there are many variations in color patterns, and alba and albescent forms are not uncommon. Read the description carefully, and if it doesn't fit, I suggest going back to the key and trying different paths, taking into consideration your plant may be an aberrant color form. Also at the end of the chapter I have listed the species by the countries in which they have been reported.

I had the good fortune to visit cultivated collections of *Kefersteinia* in Colombia. The growers had several specimens of some species and I was able to tear apart flowers and photograph the flower parts. I found certain species, especially *K. gemma* and *K. laminata*, to be very variable in size of the segments (one plant would have a short column, another a longer one, and four others were in between, and this seemed unrelated to the actual size of the flower or plant). It was tempting to get excited and name each specimen a new species, but the column, wings, basic callus, and the basic color pattern were the same in each. I have concluded that there is just variability in the species and that the world will not be a better place if I create new species to accommodate the small differences in these specimens. Other authors have created some species with what seems to be minimal criteria differentiating one from another similar species. Once the species is created, it stands, and so I will tell you how to tell the species apart, but feel there are many species in this genus that should be considered varieties or forms of other species. The text hints at which species I cannot find much difference between, and in the key I tried have make sure they fall out closely together, usually only separated by color differences.

Kefersteinia ranges from Mexico to Venezuela, Bolivia, Ecuador, and Peru. Brazil has only one species reported in its eastern region (*K. mystacina*) and I suspect the Brazilians are not looking hard enough. The plants of *Kefersteinia* occur in moist forests in shade, from close to sea level to 2000 meters in elevation.

Etymology
Named in honor of Christian Keferstein, an orchid grower of Kröllwitz and friend of Heinrich Gustav Reichenbach.

List of the Species of *Kefersteinia*

Kefersteinia alata
Kefersteinia alba
Kefersteinia andreettae
Kefersteinia angustifolia
Kefersteinia atropurpurea
Kefersteinia auriculata
Kefersteinia aurorae
Kefersteinia bengasahra
Kefersteinia benvenathar
Kefersteinia bertoldii
Kefersteinia bismarckii
Kefersteinia candida
Kefersteinia chocoensis
Kefersteinia costaricensis

Kefersteinia delcastilloi
Kefersteinia elegans
Kefersteinia endresii
Kefersteinia escalerensis
Kefersteinia escobariana
*Kefersteinia excentrica**
*Kefersteinia expansa**
Kefersteinia forcipata
Kefersteinia gemma
Kefersteinia graminea
  (type species of
  Kefersteinia)
*Kefersteinia
  guacamayoana**

Kefersteinia heideri
Kefersteinia hirtzii
Kefersteinia jarae
Kefersteinia klabochii
*Kefersteinia
  koechliniorum*
Kefersteinia lactea
Kefersteinia lafontainei
Kefersteinia laminata
Kefersteinia lehmannii
Kefersteinia licethyae
Kefersteinia lindneri
Kefersteinia lojae
*Kefersteinia maculosa**

Kefersteinia microcharis*	*Kefersteinia pseudopellita*	*Kefersteinia taggesellii*
Kefersteinia minutiflora	*Kefersteinia pulchella*	*Kefersteinia taurina*
Kefersteinia mystacina	*Kefersteinia punctatissima*	*Kefersteinia tinschertiana*
Kefersteinia niesseniae	*Kefersteinia pusilla*	*Kefersteinia tolimensis*
Kefersteinia ocellata	*Kefersteinia retanae*	*Kefersteinia trullata**
Kefersteinia orbicularis	*Kefersteinia*	*Kefersteinia urabaensis*
Kefersteinia oscarii	*richardhegerlii*	*Kefersteinia vasquezii*
Kefersteinia parvilabris	*Kefersteinia ricii*	*Kefersteinia villenae*
Kefersteinia pastorellii	*Kefersteinia sanguinolenta*	*Kefersteinia villosa*
Kefersteinia pellita	*Kefersteinia stapelioides*	*Kefersteinia vollesii*
Kefersteinia perlonga	*Kefersteinia stevensonii*	*Kefersteinia wercklei*

* Molecular sampling confirms placement of this species within the genus *Kefersteinia*.

Key to the Species of *Kefersteinia*

1a. Callus two balls or a semicircular bar, often on a stalk or post as with a mushroom stem. . . . . go to 2

1b. Callus not stalked (no stalk or stem below the callus), or if stalked, the apical side of the upper portion of the callus is more free of the lip disc than the basal portion, callus most often appearing like a butterfly with two slightly flexed wings or two tongues joined at the base, though sometimes the upper portion of the callus looks like a flat plate but is more free apically than basally. . . . . go to 20

2a. Callus a semicircular transverse bar, lip markedly fimbriate (sepals and petals green, lip white or pink). . . . . *K. mystacina*

2b. Callus two balls or globes, or a plate on a stalk. . . . . go to 3

3a. Lip as wide and long as lateral sepals are wide, lip small in relation to rest of flower. . . . . go to 4

3b. Lip wider and longer than lateral sepals are wide, lip a significant part of flower face. . . . . go to 7

4a. Lip without apical point, lateral margins and apex recurved, flower green. . . . . *K. perlonga*

4b. Lip with apical point. . . . . go to 5

5a. Callus of two keels that reflex and curl back at apex (green with purple spots). . . . . *K. endresii*

5b. Callus of a stalk and two balls at apex. . . . . go to 6

6a. Lateral edges of lip incurved, flower yellow-green without spots. . . . . *K. stevensonii*

6b. Lateral edges of lip not incurved, flower green with basal red-wine staining. . . . . *K. trullata*

7a. Callus sub-basal, basal lateral lip margins curl in, the rest of lip flat, flower yellow-green. . . . . *K. guacamayoana*

7b. Callus basal, lip otherwise. . . . . go to 8

8a. Callus without a stalk. . . . . go to 9

8b. Callus two balls on a stalk or a plate on a stalk. . . . . go to 10

9a. Lip margins fimbriate, flowers green with red-purple spots or lines. . . . . . .
. . . . *K. expansa*

9b. Lip margins not fimbriate, lip with long stiff isthmus after callus, lateral
lip edges small rolled outward (yellow-green with stripes and spots). . . . . .
. . . .*K. parvilabris*

10a. Median ventral keel tooth not extended basally more than lateral teeth,
flower densely spotted purple with green or yellow base. . . . . *K. maculosa*

10b. Median ventral keel tooth the same length as or longer than lateral teeth
of column ventral plate or lateral teeth lacking, flower mostly white or
green with few spots . . . . go to 11

11a. Lip with small basal lobes or auricles . . . . go to 12

11b. Lip without small basal lobes or auricles . . . . go to 14

12a. Lip oval in overall shape, flower white with red spots . . . . *K. auriculata*

12b. Lip widest at apex, flower green-yellow with red spots. . . . . go to 13

13a. Column with blunt lateral wings . . . .*K. ocellata*

13b. Column with two lateral acute protuberances on each side of column . . . .
. . . .*K. chocoensis*

14a. Lip widest at mid length. . . . . go to 15

14b. Lip widest at apex . . . . go to 17

15a. Lateral sepals less than 0.5 cm wide, long and thin . . . . *K. orbicularis*

15b. Lateral sepals over 0.5 cm wide. . . . . go to 16

16a. Dorsal sepal acuminate, sepals and petals green-white to yellow densely
spotted with purple. . . . . *K. costaricensis*

16b. Dorsal sepal elliptic, sepals and petals white with few scattered spots . . . .
. . . .*K. alata*

17a. Callus basal. . . . . go to 18

17b. Callus sub-basal. . . . . go to 19

18a. Ventral median keel extended to column base, ventral plate of column
square with four corners, sepals and petals spotted with red, lip white with
solid purple at apex . . . . *K. wercklei*

18b. Ventral median keel ending basally in a tooth not extended to column
base, ventral plate triangular and widest at base of plate, sepals and petals
spotted with red, lip white with purple lines at apex . . . .*K. retanae*

19a. Lip not tilted to one side, apex notched, sepals and petals spotted with red,
lip green with large rose-purple blotches . . . . *K. angustifolia*

19b. Lip tilted to one side, apex notched with central point, lip white densely
spotted with purple. . . . .*K. excentrica*

20a. Lip not deflexed sharply, not forming a showy skirt, flower generally
cupped . . . . go to 21

20b. Lip deflexed sharply, creating a skirt or flat front or face . . . . go to 28

21a. Lip extended downward 165 degrees from the column axis, flower green with purple spots. . . . .*K. licethyae*

21b. Lip otherwise . . . . go to 22

22a. Callus longer than wide with two lateral points. . . . . go to 23

22b. Callus not longer than wide . . . . go to 24

23a. Callus with one lateral point on each side, flower yellow with purple spots on petals and lip, lateral sepals held lateral to lip, not angled down . . . .*K. benvenathar*

23b. Callus with two blunt lateral points on each side, flower green. . . . . *K. minutiflora*

24a. Callus extended laterally onto or up to lateral edges of lip. . . . . . . *K. pusilla*

24b. Callus semiquadrate, not extended to lateral upturned edges of lip . . . . go to 25

25a. Lateral edges of lip with a transverse crease between base and mid length of lip. . . . . go to 26

25b. Lateral edges of lip with a transverse crease at apical third of lip. . . go to 27

26a. Lip edges denticulate all around, column with two very small acute hooked ears or wings lateral to the stigma (green with purple spots, column yellow-green) . . . .*K. bertoldii*

26b. Lip edges not denticulate, column with no ears or wings lateral to stigma (green sepals and petals, white lip with red spots, column clear green). . . . .*K. jarae*

27a. Callus widest at base and narrowing at the apex, lateral sepals longer than petals, green with red spots . . . . *K. vollesii*

27b. Callus square, lateral sepals same length as petals, green-yellow lightly spotted red-brown . . . . *K. lindneri*

28a. Callus three-lobed and fused, the lateral callus lobes square, the medial lobe twice as long and keel-like, sepals and petals green, lip pink white, column reflexes back so the apical column is vertical . . . .*K. hirtzii*

28b. Callus two-lobed only, color otherwise. . . . . go to 29

29a. Column curved on dorsum so that clinandrium on ventral surface . . . . go to 30

29b. Column straight . . . . go to 34

30a. Lip with an isthmus (long thin section) apical to the callus, narrow at base for at least one-quarter of lip length then becoming wider . . . go to 31

30b. Lip without prominent isthmus, narrow at base then quickly widening so widest at mid lip length or more apically. go to 33

31a. Column widest at stigma, wings absent, flowers green-white covered with red flecks, lip white with red spots . . . .*K. taurina*

31b. Column widest at mid column length, wings present . . . . go to 32

32a. Column foot minimal in length (0.1 cm), sepals and petals green-yellow with maroon spotting along the veins, lip with dense red spots arranged in rays apically . . . . *K. taggesellii*

32b. Column foot nearly as long as column, sepals and petals yellow with maroon spotting, denser on lip . . . .*K. lehmannii*

33a. Callus with two apical points and at least two lateral points. . . *K. graminea*

33b. Callus with two apical points and only two lateral points, one on each side . . . .*K. stapelioides*

34a. Column with no wings. . . . . go to 35

34b. Column with wings. . . . . go to 39

35a. Column with wide flat ventral surface from base of column to stigma. . . . . go to 36

35b. Column otherwise. . . . . go to 37

36a. Lip obviously notched, sepals and petals recurved. . . . . *K. vasquezii*

36b. Lip not obviously notched, sepals and petals not recurved or only slightly recurved. . . . .*K. pulchella*

37a. Column ventral surface curved slightly . . . . *K. tinschertiana*

37b. Column ventral surface straight. . . . . go to 38

38a. Column an even width for its entire length . . . . *K. tolimensis*

39b. Column with an obtuse widening mid length. . . . .*K. ricii*

39a. Lip with long isthmus at base . . . . go to 40

39b. Lip without long isthmus. . . . . go to 41

40a. Column foot 4–5 mm long, in same axis as column (not at angle to column), flower yellow . . . .*K. niesseniae*

40b. Column foot minimal . . . . *K. elegans*

41a. Flower pure or true white without spots. . . . . go to 42

41b. Flower not pure white, either different color or with spots. . . . . go to 43

42a. Flower pure white, callus without lateral points. . . . . *K. candida*

42b. Flower white with yellow on column and wings, callus with acute lateral points. . . . . *K. lafontainei*

43a. Flower buttercup yellow sometimes with small red spots. . . . . *K. andreettae*

43b. Flower not buttercup yellow . . . . go to 44

44a. Flowers white or off-white or yellow-white, with no or scattered spots on sepals and petals . . . . go to 45

44b. Flower base color densely covered by spots, so that spots are a significant portion of the flower color, especially spotted or colored on sepals and petals . . . . go to 60

45a. Callus extends apically to only one-quarter of lip length . . . . go to 46

45b. Callus extends to near lip mid length (to the transverse fold) . . . . . go to 55

46a. Column wings extend laterally not ventrally. . . . . go to 47

46b. Column wings flex ventrally . . . . go to 48

47a. Callus apexes diverge laterally, lateral callus with one notch . . . *K. urabaensis*

47b. Callus apexes not diverging but pointing apically only, lateral callus
 margin scalloped . . . .*K. klabochii*

48a. Callus with nipplelike apexes . . . . *K. escalerensis*

48b. Callus without nipplelike apexes . . . . go to 49

49a. Callus widest at base, lip disc with maroon coloration lateral to callus . . .
 . . . .*K. forcipata*

49b. Callus widest at mid length or apex, lip coloration other . . . . go to 50

50a. Lip with coloration in radiating rays from base to mid lip length . . . .
 . . . . *K. heideri*

50b. Lip with spots . . . . go to 51

51a. Flower base color pure white . . . . go to 52

51b. Flower base color yellow-white or green-yellow . . . . go to 54

52a. Lateral edges of lip complete only one-third of a circle . . . . *K. alba*

52b. Lateral edges of lip complete more than half of a circle . . . . go to 53

53a. Lip without spots . . . . *K. lactea*

53b. Lip with red spots . . . . *K. microcharis*

54a. Ventral keel of column extends to column base . . . .*K. bengasahra*

54b. Ventral keel of column is apical only, does not extend to column base . . . .
 . . . . *K. sanguinolenta*

55a. Callus with lateral points or bulges . . . . go to 56

55b. Callus without lateral points or bulges . . . . go to 58

56a. Lateral points on callus rounded . . . . *K. laminata*

56b. Lateral points on callus acute . . . . go to 57

57a. Callus with two points apically . . . . *K. gemma*

57b. Callus with four points apically . . . .*K. punctatissima*

58a. Spots pink or red . . . . *K. oscarii*

58b. Spots purple on sepals . . . . go to 59

59a. Lip white with few medial brown spots . . . . *K. delcastilloi*

59b. Lip green densely covered with purple markings . . . .*K. atropurpurea*

60a. Central column keel prominent and extending to base of column . . . .
 . . . . go to 61

60b. Central column keel not prominent and extending only to mid column
 length . . . . go to 63

61a. Central keel diminishes at midway, then thickens again and becomes rib
 that extends to the foot . . . . *K. aurorae*

61b. Central keel not diminished at mid length . . . . go to 62

62a. Callus with two keels on each side, one on top of the other . . . .
 . . . . *K. koechliniorum*

62b. Callus with only one keel on each side . . . . *K. lojae*

63a. Lip with long hairs or pubescence . . . . go to 64

63b. Lip without long hairs . . . . go to 69

64a. Lip with long villose trichomes, angle of column and its foot about
90 degrees . . . . go to 65

64b. Lip sparsely pubescent, angle of column and its foot greater than
90 degrees . . . . go to 66

65a. Lateral sepals elliptic or mildly spatulate. . . . . *K. pastorellii*

65b. Lateral sepals pear-shaped. . . . . *K. bismarckii*

66a. Sepals four times as long as wide . . . . *K. escobariana*

66b. Sepals two and one-half times as long as wide . . . . go to 67

67a. Callus without lateral auricles . . . . *K. pseudopellita*

67b. Callus with lateral auricles . . . . go to 68

68a. Callus auricles rounded . . . . *K. villosa*

68b. Callus auricles acute . . . . *K. pellita*

69a. Wings prominent at column mid length . . . . K. *villenae*

69b. No wings, only a widening mid column . . . . *K. richardhegerlii*

Kefersteinia alata Pupulin, *Harvard Pap.*
Bot. 8(2): 161. 2004. Type: Panama, Bocas del Toro, road from David to Chiriquí Grande, around km 74, 450 m, 10 April 2001, flowered in cultivation at Gaia Botanical Garden, 18 August 2001, *F. Pupulin, D. Castelfranco & E. Salas 3119* (holotype: USJ).

DESCRIPTION The flowers are borne on pendent inflorescences. The sepals are pure white, the petals are white with a few diffuse purple spots, the lip is white with dark purple spots and blotches, and the callus is white with purple spots or blotches. The sepals are concave at their apex. The petals lie in the same plane as the column. The lip is suborbicular with a short claw, the base of the lip is concave, and the apical portion recurves gently. The bilobed callus has a prominent stalk and an apical notch. The column is narrow at the base, widening at mid length but without wings. The ventral surface of the column forms a rectangular plate that protrudes basally with a central keel that becomes a central tooth; the plate also forms two shorter teeth lateral to the central tooth, all projecting basally. PLATE 57.

MEASUREMENTS Leaves 7.3–20.0 cm long, 1.6–2.8 cm wide; inflorescence 3.5 cm long with one bract; dorsal sepal 1.6 cm long, 0.6 cm wide; lateral sepals 2.1 cm long, 0.6 cm wide; petals 1.6 cm long, 0.6 cm wide; lip 1.3 cm long, 1.1 cm wide; column 0.7 cm long, 0.4 cm wide.

ETYMOLOGY Latin *alatus*, winged, in reference to the position of the petals, like birds taking flight.

DISTRIBUTION AND HABITAT Known only from Panama, at 450–1200 m in elevation along the eastern and Caribbean slopes of Barú volcano, in premontane and lower montane moist forests.

FLOWERING TIME June to September.

Kefersteinia alba Schlechter, *Repert. Spec.*
Nov. Regni Veg. Beih. 19: 228. 1923. Type: Costa Rica, Bois, San Pedro de San Ramon, 1075 m, July 1922, *A. M. Brenes 284* (lectotype designated by Barringer: AMES).

SYNONYM
Chondrorhyncha alba (Schlechter) L. O. Williams, *Ceiba* 5: 191. 1956.

DESCRIPTION The flowers are borne on pendent inflorescences and are white with a few red-brown spots on the lip. The petals have erose margins. The lip is rhomboid with a short clawed base, lightly three-lobed, and deflexes to form a skirt at its apical two-thirds length. The lip margin is subcrenulate to undulate and crisped. The basal two-lobed callus is widest apically or at mid length, smaller at its base; the callus is said to be lyre-shaped, the lobes do not bifurcate laterally at their apex. The column is bent back or slightly dorsally flexed at mid length, is widest at mid length with wings, and has a medial ventral keel below the stigma that extends to mid length of the column. PLATE 58.

COMMENT Four species of *Kefersteinia* are mostly white and have a broad skirt: *K. alba* (Costa Rica, the lateral lobes don't swing up as far as the other species, the flower is larger than *K. lactea*), *K. candida* (Peru, all white), *K. lactea* (Costa Rica, with sparse brown spots), and *K. lafontainei* (French Guiana, white with yellow on the column wings).

MEASUREMENTS Leaves 5–15 cm long, 0.6–1.5 cm wide; inflorescence 2.3 cm long, with one or more bracts; sepals 1.4–1.5 cm long, 0.4–0.5 cm wide; petals 1.3 cm long, 0.5 cm wide; lip 1.2 cm long and wide; column 0.9–1.0 cm long.

ETYMOLOGY Latin *alba*, white, describing the flower color.

DISTRIBUTION AND HABITAT Known from Costa Rica and Panama, at 300–1100 m in elevation, in semishade.

FLOWERING TIME Throughout the year.

Kefersteinia andreettae G. Gerlach, Neudecker & Seeger, *Orchidee (Hamburg)* 40(4): 133. 1989. Type: Ecuador, Cordillera del Condor, 700–1500 m (holotype: HEID).

SYNONYM
Kefersteinia salustianae D. E. Bennett & Christenson, *Brittonia* 46: 37. 1994.

DESCRIPTION The flowers are borne on pendent inflorescences. The sepals and petals are pale yellow-green to yellow. The lip is clear yellow, although some specimens have red-brown spots and blotches on the basal half of the concave sides of the lip. The callus is yellow; some specimens have scattered small reddish spots. The column is pale yellow-green, sometimes with small red spots dispersed over the ventral surface. The anther is dull yellow, and the pollinia are waxy white. The sepals are spread, the lateral sepals are concave and longer than the dorsal sepal. The petals have a thickened midvein. The obscurely three-lobed lip is concave basally. The lateral lip lobes are rounded. The midlobe is reniform, erose, and deflexes sharply at the apical quarter of the lip length. The callus is composed of three longitudinal ridges that come together at the midpoint, the middle ridge divides to become apically four-toothed. The bilobed callus is minutely tuberculate, and the apexes of the lobes have two teeth. The callus extends apically almost to the point of deflection of the lip. The column is more or less straight, convex above, more or less flat to slightly convex below, sparsely covered with a short pubescence, and has a small short median ventral keel extending basally below the stigma. The column foot is as long as the column and held at slightly more than a right angle to the column. PLATE 59.

COMMENT The column foot is unusually long for a *Kefersteinia* species, more in line with the column foot of a *Stenia*, but the callus is typical for a *Kefersteinia* and other features make it fit within that genus. Ecuagenera and others have sent me pictures of other yellow *Kefersteinia* that appear to be *K. andreettae* or a close relative. The other forms have concave lips that do not deflex downward, and some even have lips that inroll along their apical margins. If these are not hybrids, perhaps further study may determine their relationships as to species and perhaps genus.

MEASUREMENTS Leaves 16.0–17.5 cm long, 1.9 cm wide; inflorescence 6–7 cm long, with two or three bracts; sepals 1.5–1.8 cm long, 0.9–1.1 cm wide; petals 1.3–1.7 cm long, 0.9–1.1 cm wide; lip 1.1–1.4 cm long, 1.3 cm wide; column 1.3 cm long, 0.5 cm wide.

ETYMOLOGY Named for Angel Andreetta, who has been a prominent figure in orchid study in Ecuador.

DISTRIBUTION AND HABITAT Known from Ecuador and Peru, at 700–1800 m in elevation, in wet montane forest.

FLOWERING TIME Most of the year.

Kefersteinia angustifolia Pupulin & Dressler, *Harvard Pap. Bot.* 8(2): 164. 2004. Type: Panama, Chiriquí, Laguna, near El Hato del Volcán, 21 October 1967, *H. Butcher s.n.* (holotype: MO).

DESCRIPTION The flowers are borne on pendent inflorescences. The sepals and petals are pale green spotted with rose-purple, and the lip is green with large rose-purple blotches. The leaves are very narrow. The flowers are bell-shaped. The dorsal sepal is concave, and the lateral sepals are falcate and deeply concave at the base with the inner basal margin curling in. The petals are slightly concave at the apex. The three-lobed lip has a short claw and is concave along its entire length. The lateral lip lobes are obscure or small, the midlobe widens at mid length forming two erect wings, and the midlobe has a central notch apically. The bilobed callus is on a stalk and is in a sub-basal position on the lip. The column is narrow at the base, widening apically, the ventral surface has a subrectangular plate which protrudes basally into a transverse keel, and the margins of the transverse keel are lacerate. There is a protruding longitudinal keel that starts in the middle of the ventral plate and extends down the ventral surface of the column. The column foot is short.

COMMENT *Kefersteinia angustifolia* is similar to the Costa Rican *K. retanae*. The latter has a ventral keel that extends more basally and a callus that is basal. The lip of *K. angustifolia* is spatulate-flabellate, whereas the lip of *K. retanae* is obovate-pandurate. I have not found a picture of *K. angustifolia*; however, the original description has a very detailed drawing.

MEASUREMENTS Leaves 7.1–18.6 cm long, 0.5–0.7 cm wide; inflorescence 4 cm long, with two bracts; dorsal sepal 0.9 cm long, 0.4 cm wide; lateral sepals 1 cm long, 0.5 cm wide; petals 0.8 cm long, ca. 0.20–0.25 cm wide; lip 0.9 cm long, 0.8 cm wide; column 0.7 cm long, 0.3 cm wide.

ETYMOLOGY Latin *angustus*, narrow, and *folium*, leaf, referring to the narrow leaves.

DISTRIBUTION AND HABITAT Known only from Panama, at 1000 m in elevation, in lower montane moist forest.

FLOWERING TIME October.

Kefersteinia atropurpurea D. E. Bennett & Christenson, *Icon. Orch. Peruv.*, in press. Type: Peru, Junín, Chanchamayo, 10 April 1996, *O. del Castillo ex Bennett 7507* (holotype: MOL).

DESCRIPTION The flowers are borne on semi-upright inflorescences. The sepals and petals are clear green thickly overlain with dark purple transverse markings, the lip inner surface and callus are very dark purple, the column is green with dark markings halfway to its apex. The sepals and petals are ovate-elliptic. The lip deflexes to form a skirt at its apical quarter length. The bilobed callus has acute apexes, each lobe with a thick ridge which is divided basally and apically and joined in the middle. The column has two rhomboid wings at column mid length and lacks a ventral midline keel.

COMMENT When describing this species, David Bennett and Eric Christenson suggested that it might be a variety of *Kefersteinia delcastilloi*, differing in the markings, the broader column wings, and being from a much higher elevation.

MEASUREMENTS Leaves 12–15 cm long, 1.5–2.0 cm wide; inflorescence 5 cm long, with four bracts; dorsal sepal 1.6 cm long, 0.9 cm wide; lateral sepals 1.8 cm long, 1.2 cm wide; petals 1.7 cm long, 1 cm wide; lip 1 cm long, 1.2 cm wide, column 1.5 cm long, 0.5 cm wide.

ETYMOLOGY Latin *atro*, dark, and *purpureus*, purple, referring to the color of the markings of the flower.

DISTRIBUTION AND HABITAT Known only from Peru, at 2200 m in elevation, in cool wet montane forest.

FLOWERING TIME
April.

Kefersteinia auriculata Dressler,

Orquideología 16(1): 49. 1983. Type: Panama, Coclé, hills north of El Valle de Antón, prepared from cultivation, 24 July 1971, *R. L. Dressler 4063* (holotype: US; isotype: PMA).

SYNONYM
Senghasia auriculata (Dressler) Szlachetko, *J. Orchideenfr.* 10(4): 336. 2003.

DESCRIPTION The flowers are borne on semipendent inflorescences and are white. The petals and lip have maroon spots. The lip is obscurely three-lobed, the two lateral lobes are small auricles, seen best from the side, the midlobe is broadly cuneate, and the lip apex is broadly acute and somewhat reflexed. The basal bilobed callus has a stalk with ovoid callus lobes. The column is basally narrow, widest beneath the stigma, with a prominent ventral plate with median ventral keel. The plate and keel end basally in three teeth. PLATE 60.

COMMENT Robert Dressler described the species as closely resembling *Kefersteinia costaricensis*, but one can tell the difference by viewing the flower sideways to see the two basal lateral lip lobes. Dressler also notes that the callus is larger and more deeply furrowed in *K. auriculata* than in *K. costaricensis*. *Kefersteinia ocellata* is similar to *K. auriculata*; both have auricles or small lateral lobes on the lip, but the difference in color makes it easy to tell them apart.

MEASUREMENTS Leaves 4.5–15.0 cm long, 0.8–1.9 cm wide; inflorescence 1.7–2.2 cm long, with two bracts; dorsal sepal 1.1 cm long, 0.5 cm wide; lateral sepals 1.3 cm long, 0.6 cm wide; petals 0.9–1.0 cm long, 0.5 cm wide; lip 0.9–1.0 cm long, 1.1–1.2 cm wide; column 0.9–1.0 cm long.

ETYMOLOGY Latin *auriculatus*, furnished with small earlike appendages, referring to the lip.

DISTRIBUTION AND HABITAT Known only from Panama, at 800–900 m in elevation, in very wet forests.

FLOWERING TIME June to September.

Kefersteinia aurorae D. E. Bennett & Christenson, *Brittonia* 46(3): 233. 1994. Type: Peru, Junín, Chanchamayo, Kivinaki, northern slope above Río Perene, 1 September 1993, *O. del Castillo ex Bennett 5516* (holotype: NY).

DESCRIPTION The flowers are borne on pendent inflorescences. The sepals and petals can be dark yellow or green, densely marked with intense orange or dark purple-red arranged along the veins, and the lip is pale tan-yellow densely overlaid with very dark red blotches and spots and interspersed with white and dark red trichomes. The callus is pale yellow or red with dark red spots, the column is dark yellow, spotted and blotched intense brown-red, the anther is straw-yellow with rose-red spots, and the ovary is green with red-brown spots. The pubescent lip is obovate with a short claw, basally concave, deflexing at mid length sharply, the lip apex has a deeply undulate fold, and the lip margins are lacerate to fimbriate. The stalkless callus is deeply bilobed, emarginate, each lobe with a central oblong, sulcate, broadly thickened rib. The basal aspect of the callus extends laterally beyond the edges of the lateral lip margin. The column is pubescent, winged laterally below the stigma, and the medial ventral keel is short, but diminishes and then becomes a keel again at the base of the column and continues onto the column foot. The species has a musty fragrance at night. PLATE 61.

COMMENT *Kefersteinia aurorae* has a ventral column keel that diminishes somewhat at mid length but continues on down to the column foot, a trait that the other heavily pubescent *Kefersteinia* species lack; instead they have a ventral keel that extends only to mid length at maximum. *Kefersteinia aurorae* is similar to *K. pastorellii*, also from Peru. *Kefersteinia aurorae* has an elliptic dorsal sepal, while *K. pastorellii* has a pear-shaped dorsal sepal. In *K. aurorae*, the angle of the column foot to the main column is about 150 degrees, in *K. pastorellii* the angle is 90 degrees. Based on the type drawings, the columns of *K. pastorellii*, *K. pellita*, *K. pseudopellita*, and *K. villosa* are more oblong (being twice as long as wide) than the column of *K. aurorae*, which is not twice as long as wide. *Kefersteinia aurorae* lacks the lateral auricles on the callus that *K. pellita* has, and *K. aurorae* has lateral sepals that are more blunt at the apex than those of *K. pseudopellita*.

MEASUREMENTS Leaves 14.0–18.5 cm long, 1.2–1.6 cm wide; inflorescence 5.0–7.7 cm long, with two bracts; dorsal sepal 1.2 cm long, 0.6 cm wide; lateral sepals 1.5 cm long, 0.7 cm wide; petals 1.2 cm long, 0.6 wide; lip 1.3 cm long, 1.5 cm wide; column 0.9 cm long, 0.5 cm wide.

ETYMOLOGY Named to honor Aurora Pastorelli de Bennett, publisher of *Icones Orchidacearum Peruviarum*, in recognition of her care of live specimens, allowing detailed illustrations to be made.

DISTRIBUTION AND HABITAT Known from Peru, at 1700 m in elevation, at lower levels of wet cloud forest.

FLOWERING TIME September through May when in active growth.

Kefersteinia bengasahra D. E. Bennett & Christenson, *Brittonia* 46(3): 235. 1994. Type: Peru, Pasco, Oxapampa, Puerto Bermúdez, Río Lorenze, 4 June 1992. *O. del Castillo ex Bennett 5605* (holotype: MOL).

DESCRIPTION The flowers are held on semierect inflorescences. The sepals and petals are green-white, the lip is white with purple spots and flecks. The column is green-white with purple spots and flecks, the anther is yellow-white, and the pollinia are pale yellow. The flowers are bell-shaped; the sepals and petals cover the column. The lip is ovate with a claw, concave subsaccate basally, deflexing sharply at a third of its length. The apical lip margins are erose. The bilobed callus is elliptic, apically with two oblique divergent teeth, and with lateral flaps or auricles at its mid length. The stout column is slightly arcuate with wings at mid length and is sparsely pubescent dorsally and on the wings. The column has a ventral keel that becomes an oblong rib from below the middle of the column extending onto the column foot.

COMMENT Carlos Hajek has a picture of this species on his Web site of Peruvian orchids, www.peruvianorchids.org.

MEASUREMENTS Leaves 7–14 cm long, 1.2 cm wide; inflorescence 5 cm long, with two bracts; sepals and petals 1.1–1.2 cm long, 0.5 cm wide; lip 1.1 cm long, 1 cm wide; column 1.1 cm long, 0.6 cm wide.

ETYMOLOGY Named for Ben-Gasahra del Castillo, the oldest son of Oliveros del Castillo, who collected the specimen.

DISTRIBUTION AND HABITAT Known from Peru, at 1250 m in elevation, in wet montane forest.

FLOWERING TIME April to May in cultivation and probably throughout the year in the wild.

Kefersteinia benvenathar D. E. Bennett & Christenson, *Brittonia* 46(3): 235. 1994. Type: Peru, Pasco, Oxapampa, Puerto Bermúdez, Río Lorenze, 4 June 1992, *O. del Castillo ex Bennett 5606* (holotype: MOL).

DESCRIPTION The flowers are held on erect inflorescences. The sepals are translucent yellow, the petals translucent yellow with purple spots, the lip is dark purple with a contrasting pale picotee with darker spots, the callus is dark purple, the column is pale yellow, and the anther is pale yellow with several purple spots. The flowers are bell-shaped. The dorsal sepal and petals have small basal auricles. The lateral sepals are longer than the petals. The lip is obovate with a claw, obtuse, deeply concave at the base, and constricted (pinched transversely) in the middle. The basal lateral lip margins are erose and have a sparsely spiculate edge. The callus is sub-basal with two lateral teeth, two ridges in the center of the callus, and a rounded apex with two short blunt downward-projecting teeth. The callus is free except at its base and without a stalk. The column ventral surface is sparsely short-spiculate (with short hairs), and the column is widest at the stigma with a median ventral keel which diminishes gradually toward the base of the column.

COMMENT Günther Gerlach and Tilman Neudecker (1994) have this species as a synonym of *Kefersteinia pusilla*.

MEASUREMENTS Leaves 7–12 cm long, 1.2–2.0 cm wide; inflorescence 6 cm long, with two bracts; dorsal sepal 1.1 cm long, 0.4 cm wide; lateral sepals 1.2 cm long, 0.4 cm wide; petals 1.1 cm long, 0.4 cm wide; lip 1 cm long, 0.8 cm wide; column 1 cm long, 0.3 cm wide.

ETYMOLOGY Named for Ben-Venathar del Castillo, the second son of Oliveros del Castillo, the collector of the species.

DISTRIBUTION AND HABITAT Known only from Peru, at 250 m in elevation, in tropical forests.

FLOWERING TIME January through February in cultivation.

Kefersteinia bertoldii Jenny, *Orchidee (Hamburg)* 36(5): 184. 1985. Type: Peru, Amazonas, Pomacocha, 2000 m, collected by Bertold Würstle 1981, flowered in cultivation 10 April 1985, *R. Jenny 50* (holotype: G).

SYNONYM
Senghasia bertoldii (Jenny) Szlachetko & Romowicz, *Richardiana* 6(4): 182. 2006, as "*bertholdii*."

DESCRIPTION The flowers are held on erect inflorescences. The sepals and petals are pale yellow-green spotted purple and red-purple, the lip is dark purple, and the column is yellow-green with pale purple spotting or clear green. The lip is concave and lobulate with a short claw, the lateral lobe edges recurve outward, the lip is pinched transversely at mid length midway, the apical lip margins flare or decurve slightly, and all the lip margins are denticulate. The callus is transverse and extends to the lateral edges of the lip. The callus consists of two lobules with lateral posterior points. The column has a median ventral keel and two hooked ears or wings lateral to the stigma, although these wings are minimal and really appear to be more just a widening lateral to the stigma that rolls ventrally. PLATE 62.

COMMENT *Kefersteinia bertoldii* differs from *K. vollesii* in that the position of the transverse pinch of the lip is more basal in *K. bertoldii*, and the apical flare of the lip is less in *K. vollesii*. The callus of *K. bertoldii* is larger and more rounded, and the column is shorter and stouter in *K. bertoldii*. The difference between *K. jarae* and *K. bertoldii* is in the color of the lip, though this may be in the eyes of the person describing the plant. The picture of the type specimen of *K. bertoldii* by Rudolf Jenny and a series of pictures supplied to me by Carlos Uribe Vélez and showing details of *K. jarae* could easily be of the same species. If so, the name *K. bertoldii* would have priority. Günther Gerlach and Tilman Neudecker (1994) list *K. bertoldii* as a synonym of *K. pusilla*.

MEASUREMENTS Leaves 8–15 cm long, 3.0–3.5 cm wide; inflorescence 5 cm long, with two bracts; sepals 1.1–1.3 cm long, 0.7–0.9 cm wide; petals 1.1–1.3 cm long, 0.5–0.7 cm wide; lip 1.2 cm long, 1 cm wide; column 1.2 cm long.

ETYMOLOGY Named for Bertold Würstle, who first collected the species.

DISTRIBUTION AND HABITAT Known only from Peru, at 2000 m in elevation, in wet montane forest.

FLOWERING TIME January to April.

Kefersteinia bismarckii Dodson & D. E.
Bennett, *Icon. Pl. Trop*, ser. 2, 1: plate 83. 1989.
Type: Peru, Cajamarca, Santa Cruz, Hacienda
Taulis, 2000 m, collected by Paul Hutchinson
and Klaus von Bismarck, November 1964,
Hutchinson 3923-1 (holotype: UC).

DESCRIPTION The flowers are held on erect
inflorescences. The flowers are yellow-white
with small purple spots and larger purple mottling. The lip is ovate with a slight claw. The
lip base is concave with elevated lateral sides.
The lip deflexes sharply at mid length to form
a semiorbicular, semirigid ridge. The disc of
the lip is pubescent. The apical margins of the
lip are erose, and the apex of lip is retuse. The
small, basal, bilobed callus looks like two earlike structures with points apically. The column has a medial ventral keel, is without
wings, and is widest at the stigma.

COMMENT *Kefersteinia bismarckii* and *K. pastorellii* are very similar. They were both described in
the same series. They are from different locations in Peru and different elevations. From
the type drawing of each species I see these
differences: *K. bismarckii* has a greater deflection of the lip skirt (the lip skirt is at about 330
degrees from the lip disc) whereas *K. pastorellii*
has a less acute angle to the reflection (the lip
skirt is at 310 degrees from the lip disc). The
sepals of *K. bismarckii* are pyriform as compared
to the elliptic, almost spatulate sepals of *K. pastorellii*. The dilation of the lateral column edge
which could be thought of as obtuse wings is
more prominent on *K. bismarckii*.

MEASUREMENTS Leaves 20 cm long, 2 cm wide;
inflorescence 3.5 cm long with one bract; dorsal sepal 1.6 cm long, 0.5 cm wide; lateral sepals
1.5 cm long, 0.7 cm wide; petals 1.5 cm long,
0.5 cm wide; lip 1.5 cm long, 1.6 cm wide; column 1 cm long, 0.4 cm wide.

ETYMOLOGY Named for Klaus von Bismarck,
one of the collectors of the species.

DISTRIBUTION AND HABITAT Known only from
Peru, at 2000 m in elevation, in high cool wet
forest on western Andean slopes.

FLOWERING TIME January to March.

Kefersteinia candida D. E. Bennett &
Christenson, *Brittonia* 46(3): 238. 1994. Type:
Peru, Junín: Tarma, District of Huasahuasi,
Caserio Santa Rosa, 28 November 1992, *O. del
Castillo ex Bennett 5922* (holotype: MOL).

DESCRIPTION The flowers are borne on pendent inflorescences and are white. The lip is
suborbicular when expanded with a minimal
claw, the joint of lip claw and column foot is
sharply angled. The base of the lip is concave
deflexing sharply at mid length and the apical
lip is crisped and undulate with margins that
are erose and denticulate. The bilobed callus
is rhombiform, free at its apex without a stalk,
with two ridges in the center of the callus, the
lateral margins of the callus are repand (have
a slightly wavy edge). The column is winged at
mid length with a median ventral keel extending only to the base of the lateral wings, the
ventral keel reappears basally. The column is
sparsely pubescent in its mid length on both
column surfaces, and the anther is dorsally
tuberculate. PLATE 63.

COMMENT See Comment under *Kefersteinia alba*
for how to distinguish *K. candida* from the three
other white-flowered *Kefersteinia* species.

MEASUREMENTS Leaves 11–17 cm long, 1.0–
1.5 cm wide; inflorescence 4 cm long with one
bract; sepals 1.3–1.5 cm long, 0.6 cm wide; petals 1.3 cm long, 0.8 cm wide; lip 1.3 cm long,
1.5 cm wide; column 0.8 cm long.

ETYMOLOGY Latin *candida*, pure white, referring
to the flower color.

DISTRIBUTION AND HABITAT Known only from
Peru, at 780–1800 m in elevation, in wet montane forest.

FLOWERING TIME Intermittently throughout
the year.

Kefersteinia chocoensis G. Gerlach &
Senghas, *Orchidee (Hamburg)* 41(2): 45. 1990.
Type: Colombia, Chocó, 100 m, collected
by G. Seeger, flowered in cultivation in
Germany, *Botanischer Garten Heidelberg sub
0-19172a* (holotype: HEID).

SYNONYM

Senghasia chocoensis (G. Gerlach & Senghas)
Szlachetko, *J. Orchideenfr.* 10: 336. 2003.

DESCRIPTION The flowers are held on semierect
inflorescences. The flowers are white with small
red-purple spots on the lip and column. The
lip is obpyriform, the apical margin is undulate and concave, and the lateral edges fold
inward on the apical half of the lip in a pinched
appearance. There are small lateral lobes or
auricles on the basal lip. The callus consists
of two fleshy thick lobes on a stalk. The column has a narrow base, and at mid length
there are laterally two protuberances and then
another just lateral to the stigma. There is a lateral notch apical to the lateral protuberance at
the stigma, and there is a prominent median
ventral column keel. (The column could be
thought of as having an obscure ventral plate
below the stigma.) PLATE 64.

COMMENT *Kefersteinia chocoensis* is very similar in appearance to *K. ocellata*, differing in the
scalloped margins of the lip and in the column.

MEASUREMENTS Leaves 10–12 cm long, 1.8
cm wide; inflorescence 4–5 cm long, with five
bracts; sepals 0.8–0.9 cm long, 0.4 cm wide; petals 0.7 cm long, 0.3 cm wide; lip 0.7 cm long
and wide; column 0.5 cm long.

ETYMOLOGY Named for the region of Colombia
where the species was first discovered.

DISTRIBUTION AND HABITAT Known only from
Colombia, at 100 m in elevation.

FLOWERING TIME Not recorded.

Kefersteinia costaricensis Schlechter, *Beih. Bot. Centralbl.* 36(2): 413. 1918. Type: Costa Rica (holotype: B, destroyed; lectotype drawing: AMES).

SYNONYMS

Chondrorhyncha costaricensis (Schlechter) P. H.
Allen, *Ann. Missouri Bot. Gard.* 36: 86. 1949.
Senghasia costaricensis (Schlechter) Szlachetko,
J. Orchideenfr. 10: 336. 2003.

DESCRIPTION The inflorescences are pendent
to semierect. The sepals and petals are white
or light yellow, the petals have maroon spots,
and the lip is white or light yellow spotted with
maroon. The one-lobed lip is ovate with a small
claw, minimally deflexed at mid length, and
the lip apex is entire or with an obtuse apicule.
The callus is two-lobed with a stalk. The column is widest at the stigma without wings, and
there is a ventral plate below the stigma with a
median ventral keel that extends to the base of
the plate with a point. PLATE 65.

MEASUREMENTS Plant 12–18 cm tall; leaves
10–18 cm long, 1.5–2.0 cm wide; inflorescence
2.5–4.0 cm long, with two bracts; dorsal sepal
0.8–1.1 cm long, 0.4–0.5 cm wide; lateral sepals
1.0–1.3 cm long, 0.5–0.7 cm wide; petals 0.7–1.1
cm long, 0.4–0.6 cm wide; lip 0.8–1.3 cm long,
0.7–1.0 cm wide; column 0.8 cm long.

ETYMOLOGY Named for Costa Rica, the county
of original discovery.

DISTRIBUTION AND HABITAT Known from
Nicaragua to Panama, at 300–1350 m in elevation, in cloud forests.

FLOWERING TIME May.

Kefersteinia delcastilloi D. E. Bennett & Christenson, *Brittonia* 46: 32. 1994. Type: Peru, Junín, Chanchamayo, above Puerto Yurinaki, 1050 m, 7 July 1991, *O. del Castillo ex Bennett 5113* (holotype: NY).

DESCRIPTION The flowers are held on erect
inflorescences and are light green with dark
purple to brown-red spots with the spots
more densely concentrated on the dorsal sepal
and petals. The lip is white tinged green with
sparsely occurring brown spots near the middle of the inner margins, the column is very
pale green, and the anther is pale light yellow to nearly white. The lip is ovate when
expanded, shortly clawed, concave at the base,
and strongly deflexed at the apical quarter.
The basal lateral margins of the lip strongly
incurve, and the apical lip margin is pleated.

The callus extends from the base to the middle of the lip, consisting of a central low rounded elongate rib, thicker in the basal half, flanked by two erect, compressed, semiovate lobes. The callus is attenuated toward the apex. The column is arcuate forward with wings lateral to the stigma and a prominent median ventral keel below the stigma.

COMMENT The lip of *Kefersteinia delcastilloi* reflexes abruptly at the apical quarter similar to the lips of *K. gemma*, *K. laminata*, and *K oscarii*, but is oval when laid out flat compared with the more pear-shaped lips of the other species. The callus of *K. delcastilloi* lacks the lateral points of *K. gemma* and *K. laminata*. The column wings of *K. delcastilloi* are held more outward rather than pointing downward as in *K. oscarii*.

MEASUREMENTS Leaves 12–14 cm long, 1.6 cm wide; inflorescence 5 cm long, with two bracts; dorsal sepal 1.7 cm long, 0.8 cm wide; lateral sepals 2.2 cm long, 0.8 cm wide; petals 2.2 cm long, 1.2 cm wide; lip 1.5 cm long, 1.2 cm wide; column 1.4 cm long, 0.7 cm wide.

ETYMOLOGY Named for Oliveros del Castillo, who collected the species.

DISTRIBUTION AND HABITAT Known only from Peru, at 1050 m in elevation, in wet montane forest.

FLOWERING TIME January to March.

Kefersteinia elegans Garay, *Orquideología* 4: 80. 1969. Type: Colombia, Antioquia, *Gilberto Escobar 507* (holotype: AMES).

DESCRIPTION The flowers are borne on pendent inflorescences and are yellow with the lip a deeper shade of yellow. The sepals and petals are erect and do not form a hood. The lip is three-lobed and has a definite isthmus after the callus. The lip blade is sharply decurved at mid length through the lateral lobes, the middle lobe margins are undulate and slightly crispate. The callus is sub-basal and forms a vertically shaped V apically, with the ends of the V easily

seen from the front of the flower. The column has prominent elliptic wings at mid column length and a toothlike median ventral keel just below the stigma. PLATE 66.

COMMENT *Kefersteinia elegans* is similar to *K. niesseniae*; both are from Colombia. *Kefersteinia elegans* has lateral lip edges that curl, making the fold crest about three-quarters of a circle whereas *K. niesseniae* has lateral lip edges that are merely erect, forming only half a circle. The callus of *K. niesseniae* is more square-shaped overall, whereas the callus of *K. elegans* is broader at the base tapering to two points.

MEASUREMENTS Leaves 20 cm long, 3 cm wide; inflorescence 4–5 cm long; dorsal sepal 2.5 cm long, 0.7 cm wide; lateral sepals 2.8 cm long, 0.6 cm wide; petals 2.3 cm long, 0.8 cm wide; lip 2.3 cm long, 2 cm wide; column 1.2 cm long.

ETYMOLOGY Latin *elegans*, elegant, referring to the flower.

DISTRIBUTION AND HABITAT Known from Colombia and Panama, elevation not recorded.

FLOWERING TIME Not recorded.

Kefersteinia endresii Pupulin, *Ann. Naturhist. Mus. Wien*, ser. B, 103: 543. 2001. Type: Costa Rica, *A. Endrés s.n.* (holotype: W, *Herb. Reichenbach 18048*).

DESCRIPTION The flowers are borne on pendent inflorescences. The sepals are pale translucent green-white, the petals are sparsely spotted with purple, the lip is dark, maybe purple, and the callus is darker. The lip is small with a short claw, obscurely three-lobed, and pointed at the apex. The callus is sub-basal; the stalk is composed of two inflated keels that abruptly reflex, curling upward and outward, then apically and medially again to form a four-lobed tip.

COMMENT The species was collected only once, and there is no other description or illustration of a live plant. The description and drawing of the type were prepared from a herbarium specimen. No other specimens are known.

MEASUREMENTS Leaves 5.4–12.0 cm long, 0.9–1.8 cm wide; inflorescence 3 cm long, with two bracts; dorsal sepal 0.8 cm long, 0.3 cm wide; lateral sepals 1 cm long, 0.3 cm wide; petals 0.75 cm long, 0.25 cm wide; lip 0.4 cm long, 0.3 cm wide; column 0.60 cm long, 0.25 cm wide.

ETYMOLOGY Named for A. R. Endrés, who preserved the type specimen.

DISTRIBUTION AND HABITAT Known from only one site in Costa Rica.

FLOWERING TIME Not recorded.

Kefersteinia escalerensis D. E. Bennett & Christenson, *Brittonia* 46: 238. 1994. Type: Peru, San Martín, above Tarapoto, in forest along road, 4 March 1988, *R. Galvez ex Bennett 4196* (holotype: MOL).

DESCRIPTION The flowers are borne on pendent inflorescences and are off-white. The petals, base of the lip, callus, and column have scattered pale rose-purple spots. The lip is suborbicular, the margins are denticulate except at the base, the lip deflexes mid length sharply forming a transverse pleat, and the disc of the lip has six low ridges formed by veins. The callus is basal, rhombic, bilobed at the apex, terminating in constricted nipplelike tips, and is granulate. The column is subauriculate (has small ears, almost like a second set of wings) at its base and has puberulent obtuse wings at mid length. The column has a median ventral keel than extends to the column foot, and the foot has a small tooth at its apex. The anther is elliptic. PLATE 67.

COMMENT *Kefersteinia escalerensis* and *K. sanguinolenta* have similar color pattern and shape, but the column of *K. escalerensis* widens laterally at the base, forming two ears, and has obtuse wings at mid length whereas the column of *K. sanguinolenta* remains one width excepting the wings at column mid length.

MEASUREMENTS Leaves 4–15 cm long, 1.0–1.4 cm wide; inflorescence 4 cm long, with two bracts; sepals 1.2 cm long, 0.6 cm wide; petals 1.3 cm long, 0.7 cm wide; lip 1.4 cm long and wide; column 0.7 cm long, 0.5 cm wide.

ETYMOLOGY Named for Cerro Escalera, the distinctive mountain behind the city of Tarapoto in northern Peru.

DISTRIBUTION AND HABITAT Known only from Peru (perhaps also Bolivia and Ecuador, see discussion under *K. sanguinolenta*), at 1000–1700 m in elevation, in wet montane forest.

FLOWERING TIME November to December.

Kefersteinia escobariana G. Gerlach & Neudecker, *Orquideología* 19(3): 46. 1994. Type: Ecuador, Napo, between Puyo and Archidona, 750 m, collected by T. Neudecker, *Botanishche Staatsammlung München* (holotype: M).

DESCRIPTION The flowers are borne on pendent inflorescences, are yellow-green covered densely with maroon-red spots, and are mildly bell-shaped. The lip is concave at the base, deflexes sharply at mid length, and is densely covered with long hairs. The apical lip margins are undulate, and the lip apex is rounded. The stalkless callus is rhomboid, the lateral callus margins incurve forming a peak, and the pointed callus apex is slightly split. The column has two lateral wings at mid length and a median ventral keel below the stigma. PLATE 68.

COMMENT In the shape of the sepals and petals, in the callus, and in the coloration of the flower, *Kefersteinia escobariana* is very similar to *K. villenae*, but *K. villenae* lacks the prominent long white trichomes or hairs of *K. escobariana*.

MEASUREMENTS Leaves 10 cm long, 1.2 cm wide; inflorescence 3 cm long with one bract; sepals and petals 1.2 cm long, 0.4 cm wide; lip 1.1 cm long, 1 cm wide; column 0.7 cm long.

ETYMOLOGY Named for Rodrigo Escobar R. of Medellín, Colombia.

DISTRIBUTION AND HABITAT Known only from Ecuador, at 800 m in elevation, in dense forest with high humidity.

FLOWERING TIME Not recorded.

Kefersteinia excentrica Dressler & D. E. Mora-Retana, *Orquídea (México)* 13(1–2): 261. 1993. Type: Costa Rica, Cartago, La Selva, road to Taús, at 1300–1400 m, 9 November 1984, *R. L. Dressler & D. E. Mora-Retana s.n.* (holotype: USJ; isotype: FLAS).

DESCRIPTION The bell-shaped flowers are borne on pendent inflorescences. The sepals are pale green with purple spots, the petals, column, and the lip are white with purple spots and specks. The midlobe of the lip has denser spots of red-purple, and the callus is white without spots. The dorsal sepal is concave. The petal margins are denticulate. The clawed three-lobed lip is fiddle-shaped with rounded suberect lateral lobes. The lip midlobe is subflabellate and pointed at the apex with denticulate margins. The lip blade has five low midline keels. The lip is normally tilted to one side; this is the only *Kefersteinia* species on record to have this feature. The sub-basal callus is bilobed with a short stalk. The column is widest at the stigma, with a plate below the stigma having a weak median ventral keel on and below the plate. This keel forms a low tooth at the rear (basal portion) of the plate; however, this tooth is scarcely larger than the corners of the plate. PLATE 69.

MEASUREMENTS Leaves 10–15 cm long, 1.1–1.5 cm wide; inflorescence 4–5 cm long, with two bracts; dorsal sepal, 1.2–1.3 cm long, 0.5–0.6 cm wide; lateral sepals 1.5–1.7 cm long, 0.5–0.6 cm wide; petals 1.2–1.4 cm long, 0.4–0.5 cm wide; lip 1.2–1.3 cm long, 1 cm wide; column 1.0–1.1 cm long, 0.4 cm wide.

ETYMOLOGY Latin *excentricus*, one-sided or placed out of center, referring to the tilted lip.

DISTRIBUTION AND HABITAT Known from Costa Rica and Panama, at 1100–1500 m in elevation, in wet premontane forests on tree trunks at 2–5 m above the surface of the soil.

FLOWERING TIME December to March.

Kefersteinia expansa Reichenbach f., *Otia Bot. Hamburg* 31. 1878. Basionym: *Zygopetalum expansum* Reichenbach f., *Gard. Chron.* 9: 168. 1878. Type: Ecuador, *Klaboch* (holotype: W).

DESCRIPTION The flowers are held on semierect inflorescences and are light green with red-purple spots and stripes. The lip is green with brown strips on the basilar half and brown blotches on the apical half. The lip is longer than wide, decurved sharply at mid length, and is lacerate on the lateral and apical margins. The sub-basal callus is bilobed with a stalk. The column lacks a median ventral keel but has a horizontal or transverse keel. PLATE 70.

COMMENT In another description of the species, Heinrich Gustav Reichenbach says that the sepals are brown and blotched with brown at the base. Like *Kefersteinia mystacina, K. expansa* has a long lacerate to fimbriate lip, but the two species have very different calluses and *K. expansa* has dark colors in its flower, whereas the flower of *K. mystacina* is white, yellow, or green.

MEASUREMENTS Leaves 8 cm long, 2 cm wide; inflorescence 3 cm long, with two bracts; dorsal sepal 1.4 cm long, 0.5 cm wide; lateral sepals 1.6 cm long, 0.5 cm wide; petals 1.3 cm long, 0.4 cm wide; lip 2 cm long, 1 cm wide; column 0.9 cm long.

ETYMOLOGY Latin *expansus*, spread out, referring to the lip shape.

DISTRIBUTION AND HABITAT Known from western Ecuador, at 450 m in elevation, in deep shade in tropical wet forest.

FLOWERING TIME Most of the year.

Kefersteinia forcipata (Reichenbach f.) P. A. Harding, *comb. nov.* Basionym: *Zygopetalum forcipatum* Reichenbach f., *Gard. Chron.*, n.s., 20: 360. 1883. Type: Origin unknown, *Hort. Shuttleworth & Carder s.n.* (holotype: W).

DESCRIPTION The flowers are held on semiupright to pendent inflorescences. The flowers are white ochre in the original description, with a few purple spots on the apex of the lip and two brick red spaces on both sides of the callus. The column has purple lines at the base. The lateral sepals are longer than the dorsal sepal by about half the length of the dorsal sepal and slightly wider, and are held horizontal to the column. The lip is ovate with a short claw, deflexing at mid length to form a skirt, with erose apical margins. (The lip is described in the original text as fimbriate but the drawing of the type shows an erose margin.) The callus is longer than wide with its widest part being the base and has an obtuse projection on each side of the lamella. The apical ends of the lamellae are narrower than the callus base, and each lamella has an acute apex. The column is widest below the stigma with mid column wings, with perhaps two sets of wings, and no ventral keel. (The herbarium drawing shows a mid ventral keel just below the stigma, but the text says there is no ventral keel.) PLATE 71.

COMMENT The preceding narrative is based on the original description by Heinrich Gustav Reichenbach and what information one can get from Reichenbach's drawing of the holotype. Reichenbach described the flower as being large and similar to *Zygopetalum gramineum*. The callus he drew of the type is very similar to that of *K. klabochii* and that group, but I also think it looks like the callus of most any of the skirt type *Kefersteinia* species.

A drawing of *Zygopetalum forcipatum* by John Day at the Royal Botanic Gardens Kew shows a plant with yellow-green sepals and petals, a lip with dark purple basal margins, and with the general shape of *Kefersteinia graminea* or *K. tolimensis*. There is, in the original description of *K. laminata*, a statement by Reichenbach that Day's drawing of *K. laminata* is a plant he (Reichenbach) saw earlier under the greenhouse name (mislabeled) as *Z. forcipatum*. This may be why these two species have been considered synonyms.

Gerlach and Neudecker (1994) have said that this species is a synonym of *Kefersteinia laminata*. The Kew World Checklist of Monocotyledons has the citation of the basionym above as an unplaced name, and the citation *Zygopetalum forcipatum* auct., *Gard. Chron.*, n.s., 24: 70. 1885, listed as a synonym of *K. laminata*.

Ecuagenera contributed Plate 71 of the present volume, showing what I have tentatively identified as *Kefersteinia forcipata*, based on the coloration at the base of the lip and comparing the shape of the flower photographed to the original type drawings. Is the specimen pictured really *K. forcipata*? Without looking at the underside of the column and the callus from above, it is hard to tell, but since no other species of *Kefersteinia* is described has having the dark maroon patches lateral to the lip callus, this is my best guess. No data are available regarding measurements, etymology, distribution and habitat, or flowering time, and Manfred Speckmaier was unable to find the type specimen in Vienna at the herbarium.

Kefersteinia gemma Reichenbach f., *Gard. Chron.* 1: 406. 1874. Type: Country of origin for the type flower is unknown. Mounted on the type sheet in Vienna is a second specimen sent by Lalinde, identified by Reichenbach as *K. gemma* that came from Medellín, Colombia. (holotype: W).

SYNONYM
Zygopetalum gemma Reichenbach f., *Gard. Chron.* 1: 406. 1874, *in nota*.

DESCRIPTION The flowers are held on upright to pendent inflorescences. The sepals and petals are green-white with red spots or flecks; the lip is white with numerous small dark pink blotches centrally. The rhomboid lip deflexes at the apex of the callus to form a showy skirt.

The lip deflexes to the degree of almost doubling on itself. The lip margin is denticulate and crisped with a notch at the apex. The callus has two lobes with two teeth apically and with two additional small points or teeth laterally. The column is pubescent on the ventral surface. PLATE 72.

COMMENT Several *Kefersteinia* species have a lip that reflexes more than the typical 90 degrees of most skirt type species. The lip in this group angles upward from the lip base and then folds almost completely on itself. The flowers are small with round lips under 2 cm in diameter, and they range from all yellow to yellow-green with pink spots, the spots more dense and almost confluent on parts of the lip. I have tried to keep these species close together in the key, but because the only way to tell many of these species apart is by actually seeing the callus and column, be aware that telling them apart in photographs is difficult and probably it is unwise to try. Luckily, I have torn enough flowers apart to know I have correctly identified the pictured species in this book.

In photographing specimens in various collections, I have found many that fit the parameters of this species but with variable flower parts, such as a short squatty column versus a long thin column or ovate petals versus oblong petals. I am of the opinion that *Kefersteinia gemma*, along with *K. laminata* and *K. oscarii* (and perhaps a few others, see the key at beginning of the chapter), is more a species complex with variations around a central theme of small yellow-green flowers with pink, red, or purple spots, spots more dense on the lip, a lip that reflexes on itself rather severely, a two-lobed callus with variable to no lateral points, and a column with wings at mid length having a variable midline keel or tooth. One could get ambitious and give names to all the intermediaries of this group, but I am not sure that would serve a purpose. It is important to known that all specimens of this species will not look alike. See the end of the chapter for a list of those species with the type of lip shape that I call the

Gemma Complex as *K. gemma* is probably the most common species.

Dodson's plate 121 of *Kefersteinia laminata* in the *Icones Plantarum Tropicarum* (1980, ser. 1) is most likely *K. gemma*. His plate 120 of *K. gemma* in the same publication shows a flower with sepals and petals like those of *K. gemma*, but a lip that is bell-shaped, narrow, and held erect, with only the lip tip extending past the column end and only the very edge of the lip reflexing back and the outer margin of the lip rolling inward. The *Illustrated Encyclopedia of Orchids* (Pridgeon 1992) has a photograph that could be a match for the species depicted in Dodson's plate 120. I believe this could be a new undescribed species and not *K. gemma*, or it could be a form of *K. gemma* with a markedly narrow lip.

Orquídeas natives del Táchira (Fernández 2003) has a photograph labeled *Kefersteinia tolimensis*, clearly mislabeled, that could be *K. gemma* or one of that group. If so, it would extend the range of this species into Venezuela.

Plate 72 in the present volume is of an exceptionally dark colored form of *Kefersteinia gemma* and has flowers that appear very similar to flowers I have seen of *K. oscarii*; however, the flower in Plate 80 of the present volume has no acute lateral points to the callus, and the flower in Plate 72 has definite acute points. The "normal" color of *K. gemma* is more like that shown in Plate 80 of *K. laminata*. These photographs point out that you can't tell these species apart without looking at the callus. They also make one wonder how the callus varies in populations in the wild, and beg the question whether *K. laminata* and *K. oscarii* shouldn't be synonyms of *K. gemma*.

MEASUREMENTS Leaves 10 cm long, 2 cm wide; inflorescence 6 cm long, with two bracts; dorsal sepal 1 cm long, 0.4 cm wide; lateral sepals 1.2 cm long, 0.4 cm wide; petals 1 cm long, 0.4 cm wide; lip 1 cm long, 1.1 cm wide; column 0.7 cm long.

ETYMOLOGY Latin *gemma*, bud, referring to the small size of the flowers and their shape.

DISTRIBUTION AND HABITAT Known from Colombia, Ecuador, and possibly Venezuela, at 600–900 m in elevation, in deep shade in wet montane forest.

FLOWERING TIME Most of the year.

Kefersteinia graminea (Lindley)

Reichenbach f., *Bot. Zeit. (Berlin)* 10: 634. 1852. Basionym: *Zygopetalum gramineum* Lindley, *Bot. Reg.* 30: Misc. 101. 1844. Type: Colombia, Popayán, collected by Hartweg (holotype: KEW); type species of *Kefersteinia*.

SYNONYM
Huntleya fimbriata hort. Hamburg ex Reichenbach f., *Bot. Zeit. (Berlin)* 10: 634. 1852.

DESCRIPTION The flowers are borne on semi-upright inflorescences. The sepals and petals are yellow-green with red-purple markings arranged along the veins, the lip is very pale green variably covered with dark maroon-red markings that become confluent at the middle of the apex of the lip, the callus is dark maroon, and the column is green white. The oval lip is concave at the base, deflexing sharply at mid length, with small teeth and fimbriations along the apical margin, and an apical notch. The basal callus is wider than long, raised and free apically without a stalk, concave with roughly four major points apically, though the lateral edge is irregular with other minor lateral points. The callus has a finely glandular surface almost tending to glandular hairs in the center. The apically curved or rounded column becomes widest at the stigma and again at mid length though there are no wings or minimal obtuse wings, and there is a median ventral keel that forms a tooth at mid column length. The column foot is one-third the length of the total column. PLATE 73.

COMMENT Some confusion comes from the fact that the type drawings of *Kefersteinia graminea*, *K. sanguinolenta*, and *K. stapelioides* in *Xenia Orchidacea* (Reichenbach 1858–1900) have, on the same drawing, two lips labeled as *K. graminea*. This second lip is more typical of the lip and callus of *K. sanguinolenta* or perhaps *K. escalerensis*.

Kefersteinia graminea resembles *K. tolimensis* but can be distinguished by its rounded or curved apical column, whereas *K. tolimensis* has a straight column. *Kefersteinia graminea* can also be distinguished from *K. sanguinolenta*; the latter has a smaller (about half the size) lip and definite wings on the column. While the apically curved column of *K. graminea* is also seen in *K. taurina*, *K. taggesellii*, and *K. lehmannii*, the latter three have a markedly elevated callus that appears stalklike, whereas *K. graminea* has a callus that is more lobular or auricular, similar to that of *K. tolimensis*.

MEASUREMENTS Leaves 12–36 cm long, 1.2–2.5 cm wide; inflorescence 4.5–7.5 cm long with one bract; dorsal sepal 2 cm long, 1 cm wide; lateral sepals 2.3 cm long, 0.9 cm wide; petals 2 cm long, 0.9 cm wide; lip 2.1 cm long, 2.4 cm wide; column 1 cm long.

ETYMOLOGY Latin *gramineus*, grasslike, referring to the foliage.

DISTRIBUTION AND HABITAT Known from Colombia, Ecuador, and Venezuela.

FLOWERING TIME Throughout the year.

Kefersteinia guacamayoana Dodson

& Hirtz, *Icon. Pl. Trop.*, ser. 2, 6: plate 504. 1989. Type: Ecuador, Napo, Cordillera de Guacamayo, km 30 on road from Baeza to Tena, 1800 m, May 1984, *Hirtz 1726* (holotype: RPSC).

SYNONYM
Senghasia guacamayoana (Dodson & Hirtz) Szlachetko, *J. Orchideenfr.* 10: 336. 2003.

DESCRIPTION The flowers are held on semiupright inflorescences. The sepals and petals are green-yellow, the lip is yellow, and the callus a bright yellow. The long lip is truncate at its apex, slightly three-lobed at its base, and sessile. The callus is sub-basal and bilobed with a stalk. The column is slender at the base, and swollen at the apex with a median ventral keel beneath the stigma. PLATE 74.

MEASUREMENTS Leaves 14 cm long, 2 cm wide; inflorescence 4 cm long with one bract; dorsal sepal 2.3 cm long, 0.5 cm wide; lateral sepals 2.5 cm long, 0.6 cm wide; petals 2 cm long, 0.4 cm wide; lip 1.4 cm long, 1 cm wide; column 1.2 cm long.

ETYMOLOGY Named for the mountain range of Guacamayo, in eastern Ecuador.

DISTRIBUTION AND HABITAT Known from eastern Ecuador, at 1750–2000 m in elevation, in extremely wet montane forest.

FLOWERING TIME March to July.

Kefersteinia heideri Neudecker, *Orquideología* 19(3): 97. 1994. Type: Bolivia, Samaipata, 1900 m, collected by Helmut Heider, *Botanische Sammlung München* (holotype: M).

DESCRIPTION The flowers are borne on pendent inflorescences. The sepals and petals are pale green without blemishes or spots, the lip is white with rays of red centrally, and the column is green with red-maroon points arranged in lines along the ventral surface. The petals have undulate margins. The lip is concave at the base, and at mid length forms a semicircular lamina by deflexing sharply. The stalkless callus at the base of the lip is tonguelike, splitting apically with blunt ends that are free of the lip at the apex. The column is erect, widened midway with small obtuse wings at mid column length, with a median ventral keel that extends basally to the mid length of the column.

COMMENT The small flower of *Kefersteinia heideri* is very similar to those of *K. sanguinolenta* and *K. forcipata* and differs mostly in its color patterns, although the wings of *K. heideri* are obtuse and much less prominent that those of *K. sanguinolenta* and *K. forcipata*. *Kefersteinia heideri* is also similar in shape to *K. graminea*,

but the size of the flower and the calluses are different.

MEASUREMENTS Leaves 15 cm long, 0.8 cm wide; inflorescence 5 cm long, with three bracts; sepals and petals 1.6 cm long, 0.5 cm wide; lip 1.5 cm long and wide; column 1.2 cm long.

ETYMOLOGY Named for Helmut Heider of Santa Cruz, Bolivia, the discoverer of the species.

DISTRIBUTION AND HABITAT Known from Bolivia, at 1900 m in elevation.

FLOWERING TIME Not recorded.

Kefersteinia hirtzii Dodson, *Icon. Pl. Trop.*, ser. 2, 6: plate 505. 1989. Type: Ecuador, Morona-Santiago, Río Palora, 800 m, September 1983, *Hirtz 1006* (holotype: SEL).

DESCRIPTION The flowers are borne on upright inflorescences. The sepals and petals are green, and the lip is pink-white. The flower is bell-shaped with the apically concave sepals and petals surrounding the column and lip. The lip is bilobed, concave at the base, deflexing sharply for its apical quarter, and notched at the apex. The callus is three-lobed. The basal callus lobes are quadrate, and the apical lobe is an elongate keel-like lamella, 5 mm long. The column is erect, slightly bent (reflexes back so the apical column is vertical), flattened on the underside with a lateral swelling on each side at the point of the bend, and has an indistinct infrastigmatic median ventral keel. PLATE 75.

COMMENT This species certainly has an odd-shaped flower. I wonder if it could be the result of a natural hybrid between *Stenia* and *Kefersteinia*. The coloration and general shape are comparable to *Stenia calceolaris*, but the column ventral keel and lack of a significant column foot make it seem to fit into *Kefersteinia*.

MEASUREMENTS Leaves 14 cm long, 2.5 cm wide; inflorescence 6 cm long, with one or two bracts; dorsal sepal 0.9 cm long, 0.4 cm wide; lateral sepals 1.1 cm long, 0.4 cm wide; petals

1.0 cm long, 0.5 cm wide; lip 1 cm long, 1.3 cm wide; column 0.7 cm long.

ETYMOLOGY Named after Alexander Hirtz, who collected the species.

DISTRIBUTION AND HABITAT Known from eastern Ecuador, at 800 m in elevation, in extremely wet montane forest.

FLOWERING TIME June to November.

Kefersteinia jarae D. E. Bennett & Christenson, *Brittonia* 46(1): 34. 1994. Type: Peru, Huánuco, Leoncio Prado, above Tingo Maria near Las Palmas, 900 m, 10 November 1990, *E. Jara P. ex Bennett 4752* (holotype: NY).

SYNONYM
Senghasia jarae (Bennett & Christenson) Szlachetko & Romowicz, *Richardiana* 6(4): 182. 2006.

DESCRIPTION The flowers are borne on upright inflorescences. The sepals and petals are pale green-yellow with red-brown markings toward the base, the lip exterior is white, the interior has dark purple-red markings becoming more intense at the center, the dark red areas consisting of short small tuberculiform cells, the column base is clear green-yellow, the column foot is white with dark red-brown markings, and the anther is pale yellow mottled with red-brown. The slightly clawed lip is concave with the lateral margins transversely folded (pinched) at mid length and the lateral erect edges of the lip are recurved. The bilobed callus is basal to sub-basal with the lobes extending transversely to the lateral edges of lip. The callus has densely granulated minute tubercles. The column is widest at the stigma and is without wings, the dorsal lower half is sparsely pubescent, and the column has a median ventral keel that becomes a ridge dividing the concave base and foot with 8 to 10 blisters on each thickened margin. PLATE 76.

COMMENT Günther Gerlach and Tilman Neudecker (1994) have this species as a synonym of *Kefersteinia pusilla*. The difference between *K. jarae* and *K. bertoldii* is the color of the lip, though I think this may be in the eyes of the person describing the plant. The picture of the type specimen of *K. bertoldii* by Rudolf Jenny and a picture supplied to me by Carlos Uribe Vélez and showing details of *K. jarae* could easily be of the same species. The differences between these two species are slight.

MEASUREMENTS Leaves 8 cm long, 1.4 cm wide; inflorescence 3 cm long with one bract; sepals 1.2–1.4 cm long, 0.5–0.6 cm wide; petals 1.2–1.4 cm long, 0.4–0.5 cm wide; lip 1.1 cm long, 0.6 cm wide in natural position; column 0.3 cm long.

ETYMOLOGY Named for Enrique Jara P., who collected the type plant of this species.

DISTRIBUTION AND HABITAT Known from Peru, at 900 m in elevation, in montane wet forest.

FLOWERING TIME October to December and March to May.

Kefersteinia klabochii (Reichenbach f.) Schlechter, *Repert. Spec. Nov. Regni Veg. Beih.* 7: 267. 1920. Basionym: *Zygopetalum klabochii* Reichenbach f., *Gard. Chron.*, n.s., 24: 391. 1885. Type: Colombia. Syntypes: Klaboch and Dorman (holotype: W).

DESCRIPTION The sepals and petals are light green-white with a few purple spots diffusely on the sepals and on the base of the petals, the lip is white with pink or purple dots, and the column has diffuse purple spots. The lip is four-lobed, ovate, almost square, the lip margins are denticulate fringed, and the blade deflexes sharply at mid length. The bilobed callus is basal, composed of two lamellae with two apical points and two obtuse lateral points (the lateral side looks scalloped). The two lamellae diverge minimally at their apexes. The column is widest just below the stigma and has wings that are held laterally, not angled down. The ventral keel of the column is minimal, raised just off the surface of the ventral column and does not form a tooth.

COMMENT This species has been listed a synonym of *Kefersteinia laminata* from western Ecuador, which has been also considered by some authors as synonym of *K. gemma*. The Kew World Checklist of Monocotyledons lists it as an accepted species as does w³ TROPICOS. Günther Gerlach and Tilman Neudecker (1994) have this species as a synonym of *K. laminata*.

MEASUREMENTS Flower 1.3 cm long, 0.7 cm wide.

ETYMOLOGY Named for E. Klaboch, who collected the first plant sent to Heinrich Gustav Reichenbach.

DISTRIBUTION AND HABITAT Known only from Colombia.

FLOWERING TIME Not recorded.

Kefersteinia koechliniorum Christenson, *Orch. Digest* 64: 139. 2000, as "*koechlinorum.*" Type: Peru, Department of Cuzco, Aguas Calientes, Machu Picchu Pueblo Hotel, 2000 m, 25 May 1999, *M. Quispe s.n.* (holotype: NY; isotype: K).

DESCRIPTION The flowers are on somewhat upright inflorescences. The sepals and petals are translucent pale brown striped with reddish brown for their length, and are suffused with red-brown toward the apexes. The lip is similarly colored with solid dark red-brown center. It is obscurely three-lobed, emarginate, and finely pubescent. The apical lobe is small, reniform with anchoriform lateral margins. The lip reflexes severely at mid length and is minutely and irregularly denticulate at the margins. The callus is sub-basal, rhombic, emarginate, with high arching lateral margins at its base. The column is stout with broad rounded wings at mid length, is glabrous, and has a median ventral keel that extends to the base of the column. The viscidium is long and elliptical. PLATE 77.

COMMENT Other species of *Kefersteinia* have yellow-green to pale brown sepals and petals that are striped along their veins or at least have the spots arranged neatly along the veins, but *K. koechliniorum* has slightly smaller flowers and severely reflexes its lip skirt similar to the severe reflection of *K. gemma*.

MEASUREMENTS Leaves 22.5 cm long, 1.8 cm wide; inflorescence 3 cm long, with three bracts; dorsal sepal 1.3 cm long, 0.6 cm wide; lateral sepals 2.1 cm long, 0.5 cm wide; petals 1.9 cm long, 0.7 cm wide; lip 1.4 cm long and wide; column 1 cm long, 0.5 cm wide.

ETYMOLOGY Named for José and Denise Koechlin, proprietors of the Machu Picchu Pueblo Hotel in Aguas Calientes, on the grounds of which the original plant was found.

DISTRIBUTION AND HABITAT Known only to southern Peru, at 2000 m in elevation, in seasonally dry montane forest.

FLOWERING TIME September to November and March to May.

Kefersteinia lactea Reichenbach f., *Gard. Chron.* 1290. 1872, *in syn.* Basionym: *Zygopetalum lacteum* Reichenbach f., *Gard. Chron.* 1290. 1872. Type: Costa Rica. Lectotype of basionym: *Endrés 774* (lectotype: W).

SYNONYMS
Chondrorhyncha lactea (Reichenbach f.) L. O. Williams, *Caldasia* 5: 16. 1942.
Senghasia lactea (Reichenbach f.) Szlachetko & Romowicz, *Richardiana* 6(4): 182. 2006.

DESCRIPTION The flowers are produced on pendent to semiupright inflorescences. The flowers are white with a few brown-red spots and streaks at the base of the lip, sepals, and petals. The lip is suborbicular, contracted at the base into a short claw which is articulated with the base of the column. The lip deflexes sharply at mid length. The apex of the lip is entire and shallowly emarginate or with an obtuse apicule. Six clear veins extend midline from the callus to the lip apex. The callus is bilobed without a stalk and the lobes bifurcate laterally at their apex. The type drawing shows a semiterete column with a slight widening at mid length, which is most probably wings, and a median ventral keel below the stigma. PLATE 78.

COMMENT If you compare the lectotype drawings of *Kefersteinia lactea* and *K. alba* (there is no plant material of either species as a type specimen), the plants are very similar except the callus and the shape of the lip; both differences could be artistic interpretation. In the present volume, Plate 58, showing *K. alba*, and Plate 78, showing *K. lactea*, present different lip shapes, which match the type drawings. In determining the species, the shape of the lip and the size of the flower parts must be considered. See Comment under *K. alba* for how to distinguish *K. lactea* from the three other white-flowered *Kefersteinia* species.

MEASUREMENTS Leaves 6–12 cm long, 1.2–2.0 cm wide; inflorescence 1.5–2.0 cm long with one bract; sepals 0.8–1.0 cm long, 0.3–0.4 cm wide; petals 0.7–0.9 cm long, 0.3–0.4 cm wide; lip 0.6–1.0 cm long and wide; column 0.5–0.6 cm long.

ETYMOLOGY Latin *lacteus*, milky white, referring to the flower color.

DISTRIBUTION AND HABITAT Known from Mexico to Panama, at 500–1300 m in elevation.

FLOWERING TIME April to October.

Kefersteinia lafontainei Senghas & G. Gerlach, *Orchidee (Hamburg)* 41(2): 47. 1990. Type: French Guiana, south of Orapu River, south of Cayenne, 50 m, collected by A. Lafontaine (holotype: HEID).

DESCRIPTION The flowers are held on semierect inflorescences and are white with a slight yellow tinge to the wings on the column. The sepals and petals have acute apexes. The lip is ovate, deflexing sharply at lip mid length, the margins are slightly crenulate and wavy. The basal callus is raised apically without a stalk, and is concave with acute apexes and acute and long lateral points. The column is widest lateral to the stigma and also has lateral wings at mid length of the column and a median ventral keel. PLATE 79.

COMMENT *Kefersteinia lafontainei* has sepals and petals that are plainly acute, not apiculate or obtuse as in the other white *Kefersteinia* species. The transverse fold of the skirt makes the most complete circle of any of the white *Kefersteinia* species. The lip of *K. lafontainei*, when viewed from the front, looks like an uppercase U, very similar to the lips of *K. lactea* and *K. candida*. The callus of each of the three species is different. See Comment under *K. alba* for how to distinguish *K. lafontainei* from the three other white-flowered *Kefersteinia* species.

MEASUREMENTS Leaves 8–12 cm long, 0.7–1.0 cm wide; inflorescence 3 cm long, with three bracts; dorsal sepal 1.1 cm long, 0.5 cm wide; lateral sepals 1.4 cm long, 0.6 cm wide; petals 1.2 cm long, 0.6 cm wide; lip 1.3 cm long and wide; column 0.5 cm long.

ETYMOLOGY Named for André Lafontaine, who discovered the species.

DISTRIBUTION AND HABITAT Known from French Guiana, at 50 m in elevation.

FLOWERING TIME Not recorded.

Kefersteinia laminata (Reichenbach f.) Schlechter, *Repert. Spec. Nov. Regni Veg. Beih.* 7: 267. 1920. Basionym: *Zygopetalum laminatum* Reichenbach f., *Gard. Chron.*, n.s., 24: 70. 1885. Type: country unknown. Syntypes: *Hort. J. Day s.n.* and *Hort. Shuttleworth & Carder s.n.* (holotype: W).

DESCRIPTION The flowers are held on upright inflorescences. The sepals and petals are yellow-green with dark red specks, the lip is green with purple spotting, and the column is yellow with green spots. The oval or slightly elliptic lip is one-lobed with a minimal claw, sharply deflexed at mid length, and the apical margin is denticulate without a notch. The basal, bilobed callus is free at the apex without a stalk, and pointed at the apex and with lateral points. The column is short, widest mid length with lateral wings, and has a median ventral keel. PLATE 80.

COMMENT This species is very close in appearance to *Kefersteinia gemma* and *K. oscarii*, and is by some authors considered a synonym of *K. gemma*. It may be that it should be considered a variety or form of *K. gemma* given *K. laminata* has a callus with obtuse lateral points while the callus of *K. gemma* has acute lateral points. In the original description of *K. laminata* by Heinrich Gustav Reichenbach there is the statement that the drawing of *K. laminata* by John Day is a plant he (Reichenbach) saw earlier under the greenhouse name (mislabeled) as *Zygopetalum forcipatum*. This may be why these two species have been considered synonyms.

MEASUREMENTS Leaves 16 cm long, 2 cm wide; inflorescence 8 cm long with one bract; dorsal sepal 1.5 cm long, 0.5 cm wide; lateral sepals and petals 1 cm long, 0.5 cm wide; lip 1 cm long and wide; column 0.8 cm long.

ETYMOLOGY Latin *laminatus*, bladelike, a reference to the free callus or the leaves.

DISTRIBUTION AND HABITAT Known from Ecuador and Colombia, at 1000 m in elevation.

FLOWERING TIME Most of the year.

Kefersteinia lehmannii P. Ortíz,
Orquideología 20: 234. 1996. Type: Illustration by F. C. Lehmann, *Icon. Pl. Trop.*: plate 172 (lectotype: K).

DESCRIPTION The flowers are held on upright inflorescences and are white or green or light yellow, densely spotted with numerous purple dots, the dots are denser on the lip becoming solid purple centrally. The lip is long, with a long isthmus or narrow portion basally. The lip has two small lip lobules at its apex. The callus is basal, enlarged towards the apex and longi-

tudinally split. The column has a slight curve, with a base that is straight with a long foot equal in length to the actual column. The column enlarges toward its middle and has three teeth at the middle on the ventral side with a midtooth formed by the ventral keel.

COMMENT No herbarium specimen is known that matches the drawing, nor has it been rediscovered. This species was described from a drawing. *Kefersteinia taggesellii* and *K. taurina* are very similar to this species. If *K. taggesellii* is a synonym of *K. taurina*, then it could be argued that *K. lehmannii* is also a synonym of *K. taurina*.

MEASUREMENTS Leaves 23 cm long, 2 cm wide; inflorescence 6 cm long, with two bracts; dorsal sepal 2.5 cm long, 0.6 cm wide; lateral sepals 3 cm long, 0.6 cm wide; petals 2 cm long, 0.6–0.7 cm wide; lip 3 cm long, 2.5 cm wide.

ETYMOLOGY Named to honor German orchidologist F. C. Lehmann, who lived at Popayán and drew the type illustration.

DISTRIBUTION AND HABITAT Presumed to be from Colombia or Ecuador.

FLOWERING TIME Not recorded.

Kefersteinia licethyae D. E. Bennett & Christenson, *Brittonia* 46(1): 34. 1994, as "*licethyi*." Type: Peru, Pasco, Oxapampa, Caserio Eneñas, between Villarica and Cacazu, 1200 m, 16 April 1991, *O. del Castillo ex Bennett 5054* (holotype: NY).

SYNONYM
Senghasia licethyae (D. E. Bennett & Christenson) Szlachetko & Romowicz, *Richardiana* 6(4): 182. 2006.

DESCRIPTION The bell-shaped flowers are borne on pendent inflorescences. The sepals and petals are pale green densely covered with purple spots and markings becoming more dense at the bases, the lip is basally dark purple, apically green with diffuse pale purple spots, the callus is dark purple, the column is green or green-yellow, and the anther is white-green. The sepals are concave with the dorsal sepal embracing the column. The lip extends downward 165 degrees from the column axis, has a small claw, and has lateral margins that are slightly pinched near the mid length. The curved stalkless callus is erect, hirsute, wider than long, having two lobes with apical points. The column is held erect, has a medial ventral keel extending from the stigma to the column foot, and is lightly hirsute at its base both ventrally and dorsally. PLATE 81.

COMMENT Günther Gerlach and Tilman Neudecker (1994) have this species as a synonym of *Kefersteinia pusilla*.

MEASUREMENTS Leaves 5–9 cm long, 1.5 cm wide; inflorescence 4 cm long, with one or two bracts; sepals and petals 0.7 cm long, 0.3 cm wide; lip 0.8 cm long, 0.5 cm wide; column 0.7 cm long.

ETYMOLOGY Named after Licethya.

DISTRIBUTION AND HABITAT Known only from Peru, at 1200 m in elevation, in wet montane forest.

FLOWERING TIME March and April, perhaps all year.

Kefersteinia lindneri Dodson, *Icon. Pl. Trop.*, ser. 1, 5: plate 439. 1982. Type: Ecuador, Río Negro, Baños to Puyo, 1300 m, 9 July 1980, *Dodson 10582* (holotype: SEL).

DESCRIPTION The flowers are held on upright inflorescences and are green-yellow with small red-brown spots. The flowers are bell-shaped; the dorsal sepal and petals hood over the column. The lip is trullate, with a slight claw at the base; the apical third of the lip is recurved mildly. The basal hirsute callus is quadrate and flat with a short stalk. PLATE 82.

COMMENT The Kew World Checklist of Monocotyledons says this is a synonym of *Kefersteinia lojae*. Rudolf Schlechter's drawing of *K. lojae* shows a markedly toothed lip margin, whereas the *K. lindneri* lip margin is merely marginally crenulate; the column drawings of the two species are also different, with *K. lojae* being widest at column mid length and *K. lindneri* being widest lateral to the stigma.

In *Native Ecuadorian Orchids* (Dodson 2001, 2: 369) there is a photograph labeled *Kefersteinia lojae*, with a drawing on the opposite page labeled *K. lojae* Schlechter, but it is the same illustration as plate 439 of *Icones Plantarum Tropicarum*, the type illustration of *K. lindneri*, which is very different from the Schlechter drawing of *K. lojae*. If you assume that the lip of *K. lojae* drawn by Schlechter does not fold back to form a skirt, despite it being drawn as such, and if you ignore the fact that all the *K. lindneri*–*K. pusilla* group of species have the widest portion of their column at the stigma, you could put it into the *K. pusilla* group. In other words, in my opinion, you have to ignore many portions of the original description to get *K. lojae* into the *K. pusilla* group of species. This argument makes the photograph in *Native Ecuadorian Orchids* incorrectly labeled; I think the photograph is either of *K. pusilla* or a new species awaiting its own name.

MEASUREMENTS Leaves 7 cm long, 2 cm wide; inflorescence 3 cm long with one bract; sepals 1.2 cm long, 0.5 cm wide; petals 1.2 cm long, 0.4 cm wide; lip 1 cm long and wide; column 1.4 cm long.

ETYMOLOGY Named for A. Lindner, who forwarded the original specimen of the species for identification.

DISTRIBUTION AND HABITAT Known from eastern Ecuador, at 900–1300 m in elevation, from very wet montane forests.

FLOWERING TIME Most of the year.

Kefersteinia lojae Schlechter, *Repert. Spec. Nov. Regni Veg. Beih.* 8: 93. 1921. Type: Ecuador, Loja, *F. C. Lehmann s.n.* (Type: B, destroyed, drawing ex Mansfeld in *Repert. Spec. Nov. Regni Veg. Beih.* 57: plate 95. 1929).

SYNONYM
Chondrorhyncha lojae (Schlechter) C. Schweinfurth, *Bot. Mus. Leafl.* 11: 216. 1944.

DESCRIPTION The flowers are held on upright inflorescences. The color of the flowers was not given in the original description. The original description says the sepals and petals are five-nerved. The petal margins are undulate. The lip is rhombic-ovate, the inside surface is warty, the lip margin is densely crenulate and undulate, bent like a knee and curved at mid length, and the lip apex is obtuse with a notch. The callus is sub-basal with a short stalk, rhombic suborbicular, with small warts.

COMMENT This species is one with a considerable bit of confusion associated with its identity. The drawing of the type shows the lip bent back like *Kefersteinia gemma* and the column with obtuse wings at mid column length and a central ventral keel that extends to the short column foot. The lip is diamond-shaped, the basal lip margins are whole, and the apical margins are scalloped or crenulate. The callus is basal or sub-basal, square with an apical notch (could also be two lobes with blunt nonfused apexes). The callus extends for only one-fifth of the lip length.

I have seen specimens determined by Charles Schweinfurth at Chicago Field Museum labeled *Chondrorhyncha lojae* which match Schlechter's drawing. I think the closest to this specimen is *Kefersteinia aurorae*, though the column seems longer on the Chicago specimen. Comparing the Schlechter drawing with the type drawing of *K. koechliniorum*, they are remarkably similar. Of course, the problem with these identities is that both *K. aurorae* and *K. koechliniorum* are from Peru, not Ecuador.

In *Native Ecuadorian Orchids* (Dodson 2001, 2: 369) there is a photograph labeled *Kefersteinia. lojae*, with a drawing on the opposite page labeled *K. lojae* Schlechter, but it is the same illustration as plate 439 of *Icones Plantarum Tropicarum*, the type illustration of *K. lindneri*, which is very different from the Schlechter drawing of *K. lojae*. As previously discussed (see

Comment under *K. linderi*), an argument can be made that the photograph in *Native Ecuadorian Orchids* is incorrectly labeled. There are things that are notable about this photograph, however, the most important of which are the lateral sepals that are at least half again as long as the petals and held horizontal, not angling downward, as is the case in the other species of this *K. pusilla* group excepting *K. benvenathar* from Peru and maybe slightly in *K. vollesii* from Colombia. Perhaps the plant shown in *Native Ecuadorian Orchids* deserves its own species name, but without the flower in hand it is difficult to assess.

The Manual of Cultivated Orchid Species (Bechtel et al. 1986) has a picture labeled *Kefersteinia lojae*, which I believe is *K. pellita* or *K. pseudopellita*.

MEASUREMENTS Leaves 20 cm long, 1.3–2.0 cm wide; inflorescence 4–5 cm long with one bract; sepals 1.5 cm long; petals same, slightly wider than sepals; lip 1.2 cm wide; column 0.7 cm long.

ETYMOLOGY Named for the town of Loja in which the species was originally discovered.

DISTRIBUTION AND HABITAT Known from Ecuador, at 1800–2000 m.

FLOWERING TIME October.

Kefersteinia maculosa Dressler,
Orquideología 16(1): 52. 1983. Type: Panama, Veraguas, ridge of Cero Arizona, above Escuela Alto de Piedra, west of Santa Fe, 5 June 1982, *R. L. Dressler 6061* (holotype: US; isotype: PMA).

SYNONYM
Senghasia maculosa (Dressler) Szlachetko, *J. Orchideenfr.* 10: 336. 2003.

DESCRIPTION The flowers are held on upright inflorescences. The sepals and petals are translucent brown-yellow to green-yellow with purple spots at the base, the lip is light yellow and strongly spotted with purple, the callus is light yellow or orange-yellow with purple spots, and the column is green-yellow with purple spots. The lip is basally cuneate with no claw, its apical margins are crisped, retuse or obtuse, and somewhat reflexed. The sub-basal callus is bilobed with a stalk; each lobe has an irregular thickening atop and near the center. The column is narrow at the base with the ventral surface forming an elevated plate that is widest beneath the stigma and that has a median plate keel, the keel and the plate ending in three to five teeth basally, the lateral teeth extending more basally than the medial tooth.

COMMENT There is a photograph of this species at www.epidendra.org/taxones/ Kefersteinia%20maculosa/index.html#.

MEASUREMENTS Leaves 4.5–7.0 cm long, 1.3–2.4 cm wide; inflorescence 1.5–2.0 cm long, with two bracts; dorsal sepal 1.0–1.3 cm long, 0.5–0.6 cm wide; lateral sepals 1.2–1.7 cm long, 0.5–0.6 cm wide; petals 1.0–1.2 cm long, 0.5–0.6 cm wide; lip 1.2–1.3 cm long, 1.1–1.4 cm wide; column 0.8–1.0 cm long, 0.5–0.6 cm wide.

ETYMOLOGY Latin *maculosus*, full of spots, referring to the coloration of the flowers.

DISTRIBUTION AND HABITAT Known from Panama, at 850–950 m in elevation.

FLOWERING TIME June–July in cultivation.

Kefersteinia microcharis Schlechter, *Repert. Spec. Nov. Regni Veg. Beih.* 19: 300. 1923. Type: Costa Rica, San Ramon, G. Acosta, 1921 (holotype: B, destroyed; AMES, drawing).

DESCRIPTION The inflorescences are pendent. The flower is white with small red spots on the lip, and the callus is spotted brown basally. The petals are slightly smaller than the sepals. The lip is subreniform or ovate, with an apical notch and a central point at the apex; the lip margin is undulate and crenulate. The bilobed callus is rhomboid. The column is glabrous and fleshy, with minimal wings that are really a wide area lateral to the stigma and with a high ventral keel under the stigma. PLATE 83.

COMMENT Rudolf Schlechter described the species as being like *Kefersteinia lactea*, but *K. microcharis* has a wider lip and is darker in color. *Kefersteinia microcharis* can be distinguished from *K. tinschertiana* by its high ventral keel in contrast to the low ventral keel and the transverse keels on the ventral column of *K. tinschertiana*. Two photos at www.neotrop.org/gallery labeled *K. microcharis* show a lovely white flower with a nice skirt that has lateral edges like *K. lactea*, except that the lip of *K. microcharis* forms a half moon not a three-quarter circle as *K. lactea* does. *Kefersteinia microcharis* also has a few pink spots on the sepals and petals, many small pink dots on the lip skirt, and pink bars and streaks on the basal lip disc. This is in contrast to the very sparse pink spots of *K. lactea*.

MEASUREMENTS Leaves 12 cm long, 1.2 cm wide; inflorescence 2.2–3.2 cm long, bracts per inflorescence not given; sepals and petals 1.2 cm long; lip 1.1 cm long, 1.3 cm wide; column 0.9 cm long.

ETYMOLOGY Latin *micro*, very small, and *chartaceus*, papery, referring to the texture of the small flowers.

DISTRIBUTION AND HABITAT Known from Costa Rica, at 700–1050 m in elevation.

FLOWERING TIME Not recorded.

Kefersteinia minutiflora Dodson, *Icon. Pl. Trop.*, ser. 1, 5: plate 440. 1982. Type: Ecuador, Pastaza, 920 m, 12 August 1978, *Dodson 6997* (holotype: SEL).

DESCRIPTION The flowers are on upright inflorescences and are green. They are bell-shaped with a dorsal sepal that hoods over the column and with petals that hood the sides of the column to form a chamber. The one-lobed lip is concave in its basal half, lightly recurved (bent downward) at the mid portion of the lip, and pointed at the apex. The callus is rectangular, bilobed at the apex, without a stalk. The column is widest below the stigma with a ventral median keel and a second widening or lateral auricle at the column base.

MEASUREMENTS Leaves 13 cm long, 2 cm wide; inflorescence 3 cm long with one bract; dorsal sepal 0.6 cm long, 0.3 cm wide; lateral sepals 0.7 cm long, 0.2 cm wide; petals 0.7 cm long, 0.3 cm wide; lip 0.6 cm long and wide; column 0.4 cm long.

ETYMOLOGY Latin *minutus*, very small, and *florus*, flowers, referring to the small size of the flowers.

DISTRIBUTION AND HABITAT Known from Ecuador, at 1000 m in elevation, in extremely wet montane forest.

FLOWERING TIME June to August.

Kefersteinia mystacina Reichenbach f., *Gard. Chron.* 1. 530. 1881. Type: Colombia (holotype: W).

SYNONYMS
Zygopetalum mystacinum Reichenbach f., *Gard. Chron.* 1. 530. 1881, *in syn.*
Kefersteinia lacerata Fowlie, *Orch. Digest* 32: 145. 1968. Type: Colombia, Putumayo, west of Mocoa, 600 m, *FRS66C55* (holotype: LASCA).

DESCRIPTION The flowers are borne on pendent inflorescences. The sepals and petals are white to pale green, the lip is white tinted pale rose at the base, the callus has small red-purple spots, the column is pale green, and anther is light yellow. Some specimens lack the red or rose colors on the callus and lip. The sepals are keeled at their apex, and the petals are lacerate at the margins. The lip is ovate to three-lobed, deeply lacerate at the margins, and reflexed softly at mid length. The basal callus is composed of two long narrow structures that reach up toward the column almost forming a circle. The column is narrowly winged and there is median ventral keel below the stigma. PLATE 84.

COMMENT *Kefersteinia lacerata* is an accepted species in the Kew World Checklist of Monocotyledons and in w³ TROPICOS. *Kefersteinia lacerata* was found in the range of *K. mystacina* and its description matches the description of *K. mystacina*, so I am considering it a synonym.

MEASUREMENTS Leaves 16–30 cm long, 2 cm wide; inflorescence 5–7 cm long with one bract; dorsal sepal 1.6 cm long, 0.5 cm wide; lateral sepals 2 cm long, 0.5 cm wide; petals 1.8 cm long, 0.6 cm wide; lip 2.5 cm long, 3 cm wide; column 1.3 cm long.

ETYMOLOGY *Mystacina*, little cloud; perhaps Reichenbach thought the lip was cloudlike.

DISTRIBUTION AND HABITAT Known from Bolivia, Brazil, Colombia, Ecuador, Panama, and Peru, at 1000–1700 m in elevation, in wet tropical forest. The species is relatively abundant in nature.

FLOWERING TIME January to April.

Kefersteinia niesseniae P. Ortíz, *Orquideología* 20: 239. 1996. Type: Colombia, Cauca, cultivated by Andrea Niessen, June 1994, *P. Ortíz 1064* (holotype: HPUJ).

DESCRIPTION The flowers are held on upright inflorescences. The flowers are yellow, the lip is a darker yellow. The lip is truncate with a tapering base (isthmus), emarginate, widest at one-third of its length where it deflexes down sharply, and the lip margins are erose, toothed, and undulate for the apical two-thirds. The sub-basal callus begins 0.5 cm from the lip base and is laminar and semiquadrate. The base of the callus is slightly raised and enlarged, the callus apex is two-toothed with a shallow apical sinus. The column is rather short and broad with two wings at each side at mid length, the median ventral keel below the stigma is tooth-shaped. PLATE 85.

COMMENT *Kefersteinia elegans* is similar to *K. niesseniae*. Both are from Colombia. *Kefersteinia elegans* has lateral lip edges that curl to make the fold crest a three-quarter circle, whereas *K. niesseniae* lateral lip edges are merely erect making only a half circle. The callus of *K. niesseniae* is also overall more square in shape, whereas the callus of *K. elegans* is broader at the base tapering to two points.

MEASUREMENTS Leaves 29 cm long, 2 cm wide; inflorescence 5–7 cm long, with three bracts; dorsal sepal 1.9 cm long, 0. 6 cm wide; lateral sepals 2.2 cm long, 0.6 cm wide; petals 2 cm long, 0.7 cm wide; lip 2 cm long, 1.8 cm wide; column 0.7 cm long.

ETYMOLOGY Named for Andrea Niessen, owner of Orquídeas del Valle, who grew the species.

DISTRIBUTION AND HABITAT Known from Colombia.

FLOWERING TIME Recorded in June.

Kefersteinia ocellata Garay, *Orquideología* 4: 83. 1969. Type: Colombia, Antioquia, *Gilberto Escobar 504* (holotype: AMES).

SYNONYMS
Senghasia ocellata (Garay) Szlachetko, *J. Orchideenfr.* 10: 336. 2003.
Kefersteinia paradoxantha Lehmann ex P. Ortíz, *Orquideología* 20(2): 245. 1996, *nom. nud.*

DESCRIPTION The flowers are held on semierect inflorescences. The sepals and petals are yellow-green with small red spots or stripes, the lip is white with red dots or brown-red markings arranged along the veins, the callus is a dark bright yellow on a white stalk, and the column is white with red dots at the margins of its ventral plate. The brown-yellow marking are variable with different specimens. The flower is bell-shaped with sepals and petals not fully spreading. The dorsal sepal hoods over the column on some specimens, on others it is upright. The lip is concave, fleshy, and three-lobed. The lateral lobes are at the base of the lip and are small and erect like ears; the middle lobe is thickly veined, is trapezoid with a truncate apex, and has a broad apical notch with a median point. The bilobed callus has a short stalk and is kidney-shaped. The column is narrow at the base, widening below the stigma to mid length to form wings, the ventral plate below the stigma has three teeth projecting basally. PLATE 86.

MEASUREMENTS Leaves 9 cm long, 0.8 cm wide; inflorescence short, with four bracts, peduncle 3 cm long; dorsal sepal 1.2 cm long, 0.5 cm wide; lateral sepals 1.5 cm long, 0.6 cm wide; petals 1 cm long, 0.4 cm wide; lip 0.9 cm long and wide; column 0.7 cm long.

ETYMOLOGY Named for the colored basal callus in the center of the flower which gives the appearance of a pair of eyes, and for the markings on the lip which resemble the markings on an ocelot.

DISTRIBUTION AND HABITAT Known from Colombia and Ecuador, at 600–1500 m in elevation.

FLOWERING TIME December to May.

Kefersteinia orbicularis Pupulin, *Lindleyana* 15(1): 25. 2000. Type: Costa Rica, San José (province), Dota, crest of Cerro Nara, 1100 m, 15 January 1999, *F. Pupulin, D. Castelfranco & L. Spadari 1170* (holotype: USJ; isotype: SEL).

DESCRIPTION The flowers are borne on pendent inflorescences and are translucent white to pale green-white. The petals have sparse purple spots at the base, the lip is white with dark purple spots arranged along the midrib and veins and toward the base, and the callus is white to yellow. The lip is orbicular (almost circular), concave at the base but not pinched, folding gently back at mid length, and the lateral margins of the lip are crisped. The stalked basal callus is kidney-shaped and bilobed with four small points—two apically and two laterally. The column ventral surface has a transverse elliptic plate with two basally protruding short teeth laterally, and the low median keel on this plate becomes an elongate tooth centrally. The surface of the plate and teeth is covered with short velutine hairs. PLATE 87.

COMMENT There is a lovely photo of *Kefersteinia orbicularis* by Franco Pupulin at www.neotrop. org/gallery.

MEASUREMENTS Leaves 6 cm long, 1.1–1.4 cm wide; inflorescence 4 cm long, with one or two bracts; dorsal sepal 1.3 cm long, ca. 0.3 cm wide; lateral sepals 1.7 cm long, 0.3 cm wide; petals 1.2 cm long, 0.4 cm wide; lip 1.3 cm long, 1.1 cm wide; column 1 cm long, 0.5 cm wide.

ETYMOLOGY Latin *orbiculatus*, circular, referring to the outline of the lip.

DISTRIBUTION AND HABITAT Known from Costa Rica and possibly Panama, at 1100 m in elevation, in premontane wet forest and cloud forest, initially found on the mossy branches in the understory of disturbed primary vegetation, growing in shade.

FLOWERING TIME Recorded in January.

Kefersteinia oscarii P. Ortíz, *Orquideología* 20: 240. 1996. Type: Colombia, Antioquia, Peque, November 1985, *P. Ortíz 1039* (holotype: HPUJ).

DESCRIPTION The flowers are held on upright inflorescences. The sepals and petals are green-yellow with pink or red-purple spots, the lip is dark pink with white margins, and the column is white with pink spots. The leaves are rather broad for their length. The dorsal sepal and petals are erect and recurved at the apex. The lip is ovate, basally concave, deflexing sharply at mid length, and the margins of the lip are erose and undulate. The callus has two oblong lobes without a stalk, the lobes are joined basally and are deeply bifurcate apically. The callus lobes extend almost to the bend in the lip. The column has two short angular wings laterally with a thin ventral keel beneath the stigma.

COMMENT *Kefersteinia oscarii* specimens in general seem to have more pink in them and a darker pink in the lip skirt than *K. gemma* and *K. laminata*, but determining the species based on darker color can result in errors. You have to look at the callus. *Kefersteinia oscarii* has a callus with no lateral points, which differentiates it from *K. gemma* and *K. laminata*.

MEASUREMENTS Leaves 11 cm long, 2 cm wide; inflorescence 4–5 cm long, with two bracts; dor-

sal sepal 1.4 cm long, 0.6 cm wide; lateral sepals 1.5 cm long, 0.65 cm wide; petals 1.5 cm long, 0.7 cm wide; lip 1.4 cm long and wide; column 0.9 cm long.

ETYMOLOGY Named for Dr. Oscar Robledo and his wife, Marta Posada de Robledo, who cultivated the species.

DISTRIBUTION AND HABITAT Known only from Colombia.

FLOWERING TIME Recorded in November.

Kefersteinia parvilabris Schlechter, *Repert. Spec. Nov. Regni Veg. Beih.* 19: 52. 1923. Type: Costa Rica, San Jeronimo, January 1922, *C. Wercklé 116* (holotype: B, destroyed; lectotype (Christenson 2006): *Wercklé 116* drawing at AMES).

SYNONYMS
Chondrorhyncha parvilabris (Schlechter) L. O. Williams, *Ceiba* 5: 195. 1956.
Kefersteinia deflexipetala Fowlie, *Orch. Digest* 30: 117. 1966.
Senghasia parvilabris (Schlechter) Szlachetko, *J. Orchideenfr.* 10: 336. 2003.

DESCRIPTION The flowers are borne on pendent inflorescences and are yellow-green to brown. The lip is yellow or white with a red suffusion and spots, pandurate, obovate-orbicular, the basal half convexly constricted (lateral margins curved under), the apical half triangular or heart-shaped flexing forward and then apically down. The basal callus is small, reniform, and basally verrucose, and the callus lacks any conspicuous stalk. (Schlechter did not mention the warts in the original description, I think because these warts are not obvious to the naked eye.) PLATE 88.

COMMENT *Kefersteinia parvilabris* f. *escobarii* Christenson 2006. Type: Colombia, *Gilberto Escobar 609* (holotype: AMES). This form has pale yellow flowers without brown pigment.

MEASUREMENTS Leaves 10–16 cm long, 1.5–2.5 wide; inflorescence 4–5 cm long with one bract;

sepals 1.5–1.7 cm long, 0.35–0.50 cm wide; petals 1.0–1.2 cm long, 0.3 cm wide; lip 1.0–1.2 cm long, 0.4–0.5 cm wide; column 1.3 cm long.

ETYMOLOGY Latin *parvus*, little or puny, and *labrum*, lip, referring to the small lip.

DISTRIBUTION AND HABITAT Known from Costa Rica to Colombia, at 1000–1600 m in elevation, found in moist overcast rainforest growing on moss-laden branches, receiving only occasional rays of sunshine.

FLOWERING TIME June to August.

Kefersteinia pastorellii Dodson & D. E. Bennett, *Icon. Pl. Trop.* ser. 2, 1: plate 85. 1989. Type: Peru, Pasco, Yaupi, 1470 m, June 1964, *D. E. & A. P. Bennett 831* (holotype: AMES).

DESCRIPTION The flowers are held on upright inflorescences. The sepals and petals are a tawny yellow with elongate gray-brown spots, the lip is mottled with purple and dark purple spots and has a white margin, and the column is ventrally white marked with purple basally and dorsally white with several basal rose spots. The dorsal sepal is pyriform, and the spatulate lateral sepals are sharply pointed. The lip is broad, deflexed sharply at mid length, and the lip disc has an area of central pubescence. The lip margins are denticulate. The callus is fleshy, prominently bilobed without a stalk, and concave in the middle. The column is dilated just below the stigma with a low ventral median keel and is dorsally pubescent.

COMMENT *Kefersteinia bismarckii*, *K. pastorellii*, and *K. pseudopellita* are very similar. I know *K. bismarckii* and *K. pastorellii* only from their type drawings, where their columns are drawn at a 90-degree angle to the column foot. On *K. pseudopellita* the angle is much greater than 90 degrees. There are also differences in the shape of the sepals, and how the sepal and petals are held: *K. pastorellii* is a cupped flower, *K. bismarckii* is less cupped, and *K. pseudopellita* holds its sepals and petals out and open.

MEASUREMENTS Leaves 20 cm long; inflorescence 4 cm long, with two bracts; dorsal sepal 1.2 cm long, 0.55 cm wide; lateral sepals 1.7 cm long, 0.55 cm wide; petals 1.5 cm long, 0.52 cm wide; lip 1.4 cm long, 1.3 cm wide; column 1 cm long.

ETYMOLOGY Named in honor of Mario Pastorelli, who contributed greatly to the knowledge of Peru's orchids by preparing plates and line drawings.

DISTRIBUTION AND HABITAT Known only from Peru, at 1500 m in elevation, in moderately wet forests.

FLOWERING TIME January to April.

Kefersteinia pellita Reichenbach f. ex Dodson & D. E. Bennett, *Icon. Pl. Trop.*, ser. 2, 1: plate 86. 1989. Type: Ecuador, Loja, 1883, *Klaboch s.n.* (holotype: W).

DESCRIPTION The flowers are held on upright inflorescences and are pale green with dark wine-purple spots and streaks. The sepals and petals are oblong-elliptic. The lip is suborbicular, transversely sharply deflexes at mid length to form a skirt, and has denticulate apical margins. The lip is covered with sparse villose trichomes. The stalkless callus consists of a pair of puberulent keels with definite acute (based on drawing of the type) lateral auricles. The column is stout and has a faint midline ventral keel that does not extend past mid column length. The column foot and column are at about 160 degrees from each other.

COMMENT When *Kefersteinia villosa* was described, the illustration of *K. pellita* (which is incorrectly labeled and is more correctly *K. pseudopellita*) in *Icones Plantarum Tropicarum* (1989, series 2, plate 86) was used as a reference (Christenson, pers. comm.), causing the authors to create a new species, *K. villosa*, instead of considering *K. villosa* a potential specimen of *K. pellita*. *Kefersteinia pellita* would have name priority if *K. villosa* is determined to be a synonym.

I have not seen a live plant of either *Kefersteinia villosa* or *K. pellita*, or a photograph that I am certain is either species. I have seen a picture of the herbarium specimen in Vienna of *K. pellita*, which could easily be *K. villosa*, except that the auricles on the callus are smaller on *K. villosa*. The differences of the angle of the callus lateral edges or the differences of the auricles on the callus may or may not be significantly different in these two species in real life. And it may be that these differences are not significant enough to justify the separation of these two species from *K. pseudopellita*. The basal lateral edges of the callus of *K. villosa* are drawn more spreading, whereas the lateral edges of the callus of *K. pseudopellita* are at right angles to the lip disc.

All three species have the column foot in almost the same plane as the column, they have a midline ventral tooth on the column that does not descend basally and wings that extend laterally from the mid column rather than angling down or ventrally, along with sharing the same villose lip and color pattern. An argument could be made that the auricles on the callus are variable and that the angle of the callus lateral edges is not significant, and if these can be shown with live specimens, then all three species could be grouped under *Kefersteinia pellita*.

MEASUREMENTS Leaves 4–10 cm long, 1.2 cm wide; inflorescence 4 cm long, with two or three bracts; sepals 1.4–1.7 cm long, 0.5–0.6 cm wide; petals 1.4–1.7 cm long, 0.4–0.5 cm wide; lip 1.4 cm long, 1.5 cm wide; column 0.8 cm long, 0.5 cm wide.

ETYMOLOGY Latin *pellucida*, translucent or skin; according to Webster, pelisse has the same derivation and is defined as a cloak made of fur, referring to the lip with prominent hairs.

DISTRIBUTION AND HABITAT Known from Ecuador and Peru.

FLOWERING TIME Sporadically throughout the year.

Kefersteinia perlonga Dressler, *Orquideología* 18: 219. 1993. Type: Colombia, Cauca, Prepresa de Anchicayá, 500 m, 11 May 1989, *R. Escobar, W. G. H. Königer et al. 4003* (holotype: MO).

SYNONYM
Senghasia perlonga (Dressler) Szlachetko, *J. Orchideenfr.* 10: 336. 2003.

DESCRIPTION The flowers are borne on pendent inflorescences. The sepals and petals are pale yellow-green, the lip is yellow with five wine-red lines, and the column is pale yellow-green. The lip is very fleshy, especially on the lateral margins, is oblong in posture and subcircular if flattened. The base of the lip is heart-shaped with a short claw, the lateral margins and the apex of the lip are broadly rounded and recurved. The callus is weakly two-parted with a tall cylindrical stalk, the apex of the stalk being widest. The column is cylindric basally, below the stigma is a two-parted plate with a three-toothed projection basally. PLATE 89.

COMMENT *Kefersteinia perlonga* is similar to *K. stevensonii* but the flower of *K. perlonga* is larger and more colorful, the apex of lip is rounded rather than acute, and the callus is longer.

MEASUREMENTS Leaves 11–17 cm long, 1.1–1.7 cm wide; inflorescence 6–7 cm long, with two bracts; dorsal sepal 2 cm long, 0.4 cm wide; lateral sepals 1.9–2.2 cm long, 0.4–0.6 cm wide; petals 1.5–1.9 cm long, 0.2 cm wide; lip 0.6 cm long, 0.5 cm wide; column 0.7–0.8 cm long.

ETYMOLOGY Latin *perlonga*, throughout the length, referring to the exaggerated length of the callus.

DISTRIBUTION AND HABITAT Known only from Colombia, at 500 m in elevation.

FLOWERING TIME Recorded as May.

Kefersteinia pseudopellita P. A. Harding, *Orquideología* 25(2). 2008. Type: Ecuador, bloomed in cultivation in Lebanon, Oregon, 5 May 2007 (holotype: US).

SYNONYM

Kefersteinia pellita sensu Dodson & D. E. Bennett, *Icon. Pl. Trop.*, ser. 2, 1: plate 86. 1989, non Reichenbach f.

DESCRIPTION The flowers are held on upright inflorescences. The sepals and petals are green-red densely covered with red-brown mottling, the margins are less densely mottled. The lip is yellow with red to red-brown spots and blotches, the column is similarly colored, and the anther is white marked with red-blue. The sepals are mucronulate. The minimally clawed lip is elliptic, unlobed, with erose ciliate margins, and is pubescent over the entire surface. The lip folds at mid length to form a semicircular skirt, and there is a notch at the lip apex. The stalkless callus consists of a pair of puberulent keels and forms a U shape with its lateral callus margins, almost to the point that the lateral edges of the callus are at right angles to the disc of the lip. The column is stout and pubescent across the wings. The column has narrow column wings at the mid length just below the stigma, and has a faint midline ventral keel that does not extend past mid column length. The viscidium is spade-shaped. PLATE 90.

COMMENT Heinrich Gustav Reichenbach drew *Kefersteinia pellita* based on the Klaboch specimen, but it was never published. In validating Reichenbach's unpublished species at the herbarium in Vienna, Calaway Dodson did not consider the lateral auricles on the callus nor the acute apexes of the callus on the type. In the original description of *K. pellita* in *Icones Plantarum Tropicarum* (1989, series 2, plate 86), Dodson and Bennett illustrated a specimen with sepals and petals that are much rounder at their apexes than the type specimen and a callus with rounded apexes and no lateral auricles. Their specimen, *D. & A. Bennett 3829* (MO), was collected in October 1987 from

Pasco, Peru, above Yaupi Bajo, at 1900 meters. If indeed this specimen 3829 has no auricles, then it would be a different species than the one in the herbarium at Vienna, and in 2008 I made the combination *K. pseudopellita*.

Kefersteinia bismarckii, *K. pastorellii*, and *K. pseudopellita* are very similar, whereas *K. pellita* and *K. villosa* are similar. See description of *K. pastorellii* for details. I did try to consider *K. pseudopellita* as a synonym of *K. aurorae*, but the column and column ratio of length to width are different, and the pattern of coloration of the sepals and petals is different. Could they all be synonyms of *K. pellita*? I need to see more live material and/or pictures to determine this.

MEASUREMENTS Leaves 16 cm long, 1.6 cm wide; inflorescence 4 cm long, with two or three bracts; dorsal sepal 1.3 cm long, 0.7 cm wide; lateral sepals 1.6 cm long, 0.7 cm wide; petals 1.5 cm long, 0.7 cm wide; lip 1.4 cm long, 1.2 cm wide; column 1 cm long, 0.6 cm wide.

ETYMOLOGY Latin *pseudo*, false, and *pellucida*, translucent or skin. According to Webster, the word *pelisse* has the same derivation (Latin) and is defined as a cloak made of fur, referring to the lip with prominent hairs.

DISTRIBUTION AND HABITAT Known from Ecuador and Peru, at 1900 m in elevation, in cool wet cloud forests.

FLOWERING TIME Sporadically throughout the year.

Kefersteinia pulchella Schlechter, *Repert. Spec. Nov. Regni Veg.* 27: 68. 1929. Type: Bolivia, La Paz, Polo Polo, near Coroico, Norte Yung, 1100 m, November 1912, *Otto Buchtein 3697* (holotype: B, destroyed; lectotype designated by Gerlach & Neudecker, 1994: AMES).

DESCRIPTION The flowers are held on erect inflorescences and are green. The lateral sepals and petals are flecked with red, the lip underside and outer margins of the inner side are pale green, the mid portion of lip, the callus, and the base of the column are red-maroon. The one-lobed lip is without a claw and deflexes sharply at mid length. The lip margins are undulate and crenulate with occasional notches, the lip surface is glandular, giving it a velvety appearance. The callus consists of two lamellae joined basally with an apical notch, the apexes diverging minimally. The column widens from the base to the stigma forming a flat plate (one could think of these as long obtuse wings that extend the length of the column, but to me they don't look winglike). There is a minimal median ventral keel to the column. PLATE 91.

COMMENT The coloration pattern of this species is the same as *Kefersteinia vasquezii*. The velvety glandular texture of the lip in *K. pulchella* is not discussed in the description of *K. vasquezii*. The drawing of the type of the species of each varies a little in minor ways and could be accounted for by artistic interpretation. I have left these two spectacularly colored and dramatic species separate, but feel that the reality is that *K. vasquezii* is a synonym, variety, or form of *K. pulchella*. The type drawing of *K. vasquezii* shows a notch at the apex of the lip, a broader, more rhomboid ventral surface plate of the column, a callus that diverges widely at its apex, and sepals and petals that recurve back at their apexes. The type drawing of *K. pulchella* shows a lip with a possible notch at its apex but if it is there it is subtle, the flat plate on the ventral surface of the column is more pear-shaped with the widest portion being at the plate's base, and the sepals and petals are not recurved.

MEASUREMENTS Leaves 17–22 cm long, 1–3 cm wide; inflorescence 4–6 cm long, with one or two bracts; dorsal sepal 1.5–2.2 cm long, 1.7 cm wide; lateral sepals 2.0–2.2 cm long, 1.7 cm wide; petals 1.5–1.8 cm long, 1 cm wide; lip 1.7 cm long, 1.6–2.2 cm wide; column 1.2 cm long.

ETYMOLOGY Latin *pulchellus*, beautiful and little, referring to the flower.

DISTRIBUTION AND HABITAT Known only from Bolivia, at 1100 m elevation.

FLOWERING TIME Recorded in October and November.

Kefersteinia punctatissima Bennett & Christenson, *Icon. Orch. Peruv.*, in press. Type: Peru, Junín, Chanchamayo, Nueva Italia, 2200 m, April 1994, *O. del Castillo ex Bennett 6500* (holotype: MOL).

DESCRIPTION The flowers are on pendent inflorescences and are brown-green densely spotted purple, with the spots on the lip coalescing. The column is clear green except that the foot has small purple spots, and the anther is yellow-white. The lip is ovate when expanded and is shortly clawed. The basal lateral margins of the lip strongly incurve. The lip disc is concave at the base, strongly deflexed at the apical one-quarter length, and the apical margin is pleated. The callus extends from the base to the middle of the lip, the two wedges of the callus consist of a medial lamella with a flatter portion laterally so that apically there are four points, and the callus wedges each have a lateral auricle on their basal portion. The column is arcuate forward with wings lateral to the stigma and a prominent median ventral keel below the stigma.

COMMENT This species is very similar to *Kefersteinia delcastilloi* and other species in the *K. gemma* complex, but differs by the size of the plant, the flowers being more densely spotted, and the callus having auricles, in addition to other subtle differences.

MEASUREMENTS Leaves 8–11 cm long, 2.0–2.1 cm wide; inflorescence 5 cm long, with two bracts; dorsal sepal 1.5 cm long, 0.75 cm wide; lateral sepals 1.6 cm long, 0.7 cm wide; petals 1.5 cm long, 0.8 cm wide; lip 1.2 cm long and wide; column 0.8 cm long, 0.5 cm wide.

ETYMOLOGY Latin *punctatus*, dotted, referring to the spots on the flowers.

DISTRIBUTION AND HABITAT Known only from Peru, at 2200 m in elevation, in wet montane forest.

FLOWERING TIME April.

Kefersteinia pusilla (C. Schweinfurth)

C. Schweinfurth, *Fieldiana, Bot.* 33: 59. 1970. Basionym: *Chondrorhyncha pusilla* C. Schweinfurth, *Amer. Orchid Soc. Bull.* 12: 384. 1944. Type: Peru, Junín, Chanchamayo Valley, 1200 m, December 1924–1927, *Carlos Schunke s.n.* (holotype: F).

DESCRIPTION The flowers are held on semierect inflorescences. The sepals are semitranslucent pale yellow with brown-red flecks and blotches on the basal two-thirds, the petals are paler yellow with fewer brown-red markings, the lip apex is pale tan-yellow and basally densely marked with dark purple-red and red hairs, the callus is dark purple-red on its lateral lobes, the central callus is pale red with short pubescent dark red spots, the column is pale yellow with small pale red spots below mid length on the ventral side, and the anther is light yellow. The lip has a deep incurved transverse constriction (pinch) just below the mid length, the lateral margins of the apical lip are minutely multiplicate with very small papillae along the edge with the lip interior having short crystalline red and hyaline pubescence that is interspersed with 2- to 3-mm-long hispid trichomes (rough and bristly hairs). The stalkless callus is transverse, emarginate, and free except at the base, with a thickened furrowed broad rib in the middle. The column has a high infrastigmatic median keel that diminishes to a raised nerve at the base of the column foot. PLATE 92.

COMMENT Günther Gerlach and Tilman Neudecker (1994) listed *Kefersteinia benvenathar*, *K. bertoldii*, *K. jarae*, *K. licethyae*, and *K. vollesii* as synonyms of *K. pusilla*. All are of the group of *Kefersteinia* species that have a pinch, or transverse indention, of the lateral edges of the lip without forming a reflexed skirt. Gerlach and Neudecker don't list *K. lojae* as a synonym of *K. pusilla*, and my feeling from looking at the type drawing of *K. lojae* by Rudolf Schlechter is that *K. lojae* and those of the *K. pusilla* group are unrelated (see discussion under *K. lojae*). Gerlach and Neudecker also do not mention *K. lindenii* in their list of synonyms, but it would be included in the *K. pusilla* group and perhaps *K. minutiflora*.

I received several photographs of specimens from this *Kefersteinia pusilla* group. I have spent many hours trying to place names to these specimens. There are subtle differences. Believe me, if I didn't think there were differences, it would have been easier to lump them all together as one species. It may be these species would be better served as varieties or forms of *K. pusilla*. In *Native Ecuadorian Orchids* (Dodson 2001, 2: 369) the photo labeled *K. lojae* is either *K. pusilla* or a new species.

MEASUREMENTS Leaves 5–12 cm long, 1.2–1.9 cm wide; inflorescence 2.0–3.5 cm long, with two or three bracts; sepals 1.1–1.4 cm long, 0.4–0.7 cm wide; petals 1.1–1.4 cm long, 0.4–0.5 cm wide; lip 1.0–1.2 cm long, 0.5 cm wide (0.9–1.0 cm wide when expanded); column 0.9 cm long, 0.35 cm wide.

ETYMOLOGY Latin *pusillus*, very small, referring to the size of the flower relative to the other species known at the time.

DISTRIBUTION AND HABITAT Known only to Peru, in the cool lower levels of the wet cloud forest.

FLOWERING TIME Most of the year.

Kefersteinia retanae G. Gerlach ex C. O.

Morales, *Brenesia* 52: 75. 1999. Type: Costa Rica, San José (province), Cantón Pérez Zeledón, Peña Blanca, 600 m, June 1992, flowered in cultivation at Munich Botanical Garden, *Günther Gerlach 69249* (holotype: USJ).

SYNONYMS
Senghasia retanae (G. Gerlach ex C. O. Morales) Szlachetko, *J. Orchideenfr.* 10: 336. 2003.
Kefersteinia retanae G. Gerlach, *Brenesia* 41–42: 100. 1994, *nom. invalid.* (no type mentioned).

DESCRIPTION The flowers are held on upright inflorescences. The sepals and petals are yellow-green, the sepals have one minor spot but the petals have intense red spotting, and the lip is white with some purple in its base transforming to lines in the center and spots on the lip apex. The three-lobed lip is short and pyriform with a short claw, flat basally, the apical lip portion, as seen from below, is triangular, and the lip apex is slightly notched. The basal stalked callus is anvil-shaped with a central groove and one superficial wart. The column has a transverse keel apically forming a ventral plate, and a longitudinal keel. PLATE 93.

COMMENT There is a photo of this species by Franco Pupulin at www.neotrop.org/gallery.

MEASUREMENTS Leaves 17 cm long, 1.2 cm wide; inflorescence 4 cm long, with four bracts; dorsal sepal 1 cm long, 0.6 cm wide; lateral sepals 1.2 cm long, 0.6 cm wide; petals 1 cm long, 0.3 cm wide; lip 1.2 cm long, 0.6 cm wide; column 0.8 cm long, 0.4 cm wide.

ETYMOLOGY Dedicated to Dora E. Mora-Retana, who was director of the Lankester Botanical Garden in Costa Rica.

DISTRIBUTION AND HABITAT Known from Costa Rica, on the Pacific side, at 600 m in elevation.

FLOWERING TIME Not recorded.

Kefersteinia richardhegerlii R. Vásquez & Dodson, *Rev. Soc. Boliv. Bot.* 3(1–2): 12. 2001. Type: Bolivia, Cochabamba, Carrasco, 400 m, 30 January 1994, *R. Vásquez & D. Ric 2169* (holotype: LPB; isotype: VASQ).

DESCRIPTION The flowers are borne on pendent inflorescences. The sepals and petals are white-green with red-purple spots, the spots being more dense at the base, and the lip is dark purple basally, apically purple with dark purple spots and a white margin, sometimes the skirt is white without spots. The fleshy lip has a definite small claw, is concave at the base, then expands at mid length and reflexes sharply, the lip margin is denticulate with an apical point. The fleshy callus is fan-shaped and com-

posed of six ridges. The column is widest at the stigma with minimal wings just below the stigma, and with an inconspicuous lamella or ridge on the ventral surface basally.

COMMENT There is a picture of this species at www.biopat.de/kat_out/kat_out1.htm.

MEASUREMENTS Leaves 7 cm long, 1 cm wide; inflorescence 4–5 cm long, with three or four bracts; sepals 1.4 cm long, 0.4 cm wide; petals 1.3 cm long, 0.3 cm wide; lip 1.4 cm long, 1 cm wide; column 1.1 cm long.

ETYMOLOGY Named for Richard Hegerl for his contribution to the investigation and conservation of the orchids of Bolivia.

DISTRIBUTION AND HABITAT Known only from Bolivia, at 400 m elevation, in the forest foothills.

FLOWERING TIME January.

Kefersteinia ricii R. Vásquez & Dodson, *Rev. Soc. Boliv. Bot.*, 2(1): 4. 1998. Type: Bolivia, Santa Cruz, Cabellero, El Tunal, northeast of Comparapa, collected and cultivated by Darwin Ric, flowered in cultivation 24 October 1990, *R. Vásquez 1261* (holotype: LPB).

DESCRIPTION The flowers are borne on semi-pendent inflorescences. The sepals and petals are white with a tint of green, the lateral sepals have dark violet flecks, the lip is white with streaks and spots of dark violet, the callus is white with a violet tint, and the column has spots of violet. The lip is rhomboid, concave, and folds midway; the lip apex tapers mildly to a point with a notch. The stalkless callus consists of two lamellae that bifurcate apically. The straight column has no wings but widens at mid length and has a midline ventral keel that does not extend to the base.

COMMENT I know this species only from the drawing with the type description. I have seen forms of *Kefersteinia tolimensis* that are smaller, have a narrower lip, yet still lack wings on the column, but have this widening at midcolumn length. I wonder if this species is one of these forms.

MEASUREMENTS Leaves 10–18 cm long, 0.5–1.0 cm wide; inflorescence 7–8 cm long, with three bracts; dorsal sepal 1.6 cm long, 0.4 cm wide; lateral sepals 1.6 cm long, 0.5 cm wide; petals 1.5 cm long, 0.4 cm wide; lip 1.7 cm long, 1.1 cm wide; column 1 cm long.

ETYMOLOGY Named for the collector and cultivator, Darwin Ric.

DISTRIBUTION AND HABITAT Known only to Bolivia, in humid montane forest.

FLOWERING TIME Recorded in October.

Kefersteinia sanguinolenta Reichenbach
f., *Bot. Zeit. (Berlin)* 10: 635. 1852. Type: Venezuela (holotype: W).

SYNONYMS
Zygopetalum sanguinolentum (Reichenbach f.) Reichenbach f., *Ann. Bot. Syst.* 6: 658. 1861, *in syn*.
Kefersteinia sanguinea Pritzel, *Ic. Ind.* 2: 161. 1855 (*sphalm.*).

DESCRIPTION The flowers are borne on semi-upright inflorescences. The sepals and petals are straw-green to green-yellow often with small blood-red spots on the lower half, the lip is lighter green to off-white with denser spots, and the column is spotted red. The lip is one-lobed to lightly four-lobed, deflexing sharply at mid length to form a skirt. The upper lip surface is covered with fine hairs for a velvety appearance. The callus has a central sulcus forming two spreading lamellae with pointed apexes. The column has wings at mid length on the column and has a prominent ventral keel from the stigma to mid length of the column. PLATE 94.

COMMENT Plate 94 in the present volume shows a typical *Kefersteinia sanguinolenta*. I have photographed and torn apart many other color forms and yet they share the same physical charac-

teristics of *K. sanguinolenta*. This is one species where to try to match the species with the color photograph will not be helpful as the species is so variable. Population studies would be very interesting to see if there is any pattern to the different color variations. A species with a color pattern and shape similar to *K. sanguinolenta* is *K. escalerensis*, but the column of *K. escalerensis* widens laterally at the base forming two ears or wings, as well as having obtuse wings at mid length whereas the column of *K. sanguinolenta* remains one width excepting the wings at mid length.

MEASUREMENTS From *Venezuelan Orchids Illustrated* (Dunsterville and Garay 1959–1976): Leaves 20 cm long, 2.5 cm wide; inflorescence 2.5 cm long with one bract; dorsal sepal and petals 1 cm long, 0.5 cm wide; lateral sepals 1.2 cm long, 0.5 cm wide; lip 1 cm long and wide; column 1 cm long, 0.4 cm wide.

ETYMOLOGY Latin *sanguineus*, blood-red, and *lentus*, an abundance of markings, describing the numerous red markings on the flower.

DISTRIBUTION AND HABITAT Known from Colombia and Venezuela, south to Bolivia, Ecuador, and Peru (?), at 800–2000 m elevation, in wet tropical forest.

FLOWERING TIME January to April.

Kefersteinia stapelioides Reichenbach f.,
Bot. Zeit. (Berlin) 10: 64. 1852. Type: Venezuela (holotype: W).

SYNONYMS
Zygopetalum moritzii Reichenbach f., *Ann. Bot. Syst.* 6: 658. 1863.
Chondrorhyncha stapelioides (Reichenbach f.) L. O. Williams, *Ceiba* 5: 195. 1956.
Kefersteinia moritzii Reichenbach f. ex Gerlach & Neudecker, *Orquideología* 19(3): 45. 1994, *nom. nud.*

DESCRIPTION The flowers are on upright inflorescences. The sepals and petals are brown-green. The lip is light yellow with purple spots, rhomboid, weakly three-lobed, with erose apical lip margins, and the apex of the lip has a notch. The stalkless basal callus is said to be kidney-shaped, but to me it appears (based on the type drawing) more flattened with two small apical points and two larger lateral points. The apically curved column is widest at half its length, with a suggestion of two obtuse wings on each side, and a midline ventral keel from the stigma tapering to the base of the column.

COMMENT The drawings of the callus vary in the literature. Sometimes it is a small bilobed triangle at the base of the lip, sometimes it is obviously depicted as larger in size relative to the size of the lip with two apical points and two lateral points. The callus is very similar to the callus of *Kefersteinia graminea*, and the two species could be considered synonyms. I tried to make many of the photographs submitted into this species, but I was always able to get them to fit better with other species.

MEASUREMENTS Dorsal sepal 1.8 cm long, 0.7 cm wide; lateral sepals 2 cm long, 0.75 cm wide; petals 1.7 cm long, 0.7 cm wide; lip 1.8 cm long, 2 cm wide; column 1.5 cm long.

ETYMOLOGY *Stapelia*, a densely spotted genus in Asclepiadaceae (milkweed family), and Greek *oides*, like, referring to the color pattern.

DISTRIBUTION AND HABITAT Recorded in Venezuela.

FLOWERING TIME Recorded in September to December and February.

Kefersteinia stevensonii Dressler,
Orquideología 7: 135. 1972. Type: Ecuador, Pichincha, Santo Domingo de los Colorados, July 1967, *F. L. Stevenson 81270-2* (holotype: US).

DESCRIPTION The flowers are borne on semi-pendent inflorescences. The sepals and petals are yellow-white to light yellow, the lip is ivory sometimes with yellow spots on the callus, and the column is yellow-green sometimes with tiny red spots. The short lip is narrowed into a claw at its base, ovate-trulliform (egg-shaped base and triangular apex), with a pointed apex. The margin of the lip is thick. The bilobed callus is subquadrate with a stalk. The column is dilated at midway with a ventral plate and a median keel that extends and projects basally as a tooth, and the rostellum is hooded by a thin U-shaped plate at the apex of the column. PLATE 95.

MEASUREMENTS Leaves 3.0–14.5 cm long, 0.8–1.5 cm wide; inflorescence 3.0–4.5 cm long, with three or four bracts; dorsal sepal 1.2–1.3 cm long, 0.3–0.4 cm wide; lateral sepals 1.3–1.4 cm long, 0.3–0.4 cm wide; petals 1.0–1.1 cm long, 0.3 cm wide; lip 0.7 cm long, 0.5 cm wide; column 0.5 cm long.

ETYMOLOGY Named for F. L. Stevenson, of Georgia, who collected this species.

DISTRIBUTION AND HABITAT Known from western Ecuador, recorded at 300–600 m in elevation, in shade in montane cloud forest.

FLOWERING TIME Most of the year.

Kefersteinia taggesellii Neudecker,
Orquideología 19(3): 98. 1994. Type: Colombia, without exact location, flowered in cultivation in Germany, 1 August 1994, *G. Krönlein Botanishe Sammlung München* (holotype: M).

DESCRIPTION The flowers are borne on semi-erect inflorescences. The sepals and petals are a pallid green-yellow with maroon spotting along the veins, the lip at the base is red-black becoming dense red spots arranged in rays apically, and the column is green with red dots at the base. The lateral sepals reflex back. The one-lobed lip has a long claw which forms an isthmus before the lip widens to the skirt. The base of the lip is concave, deflexing sharply to form a semicircular skirt. The apical lip margins are erose or teethlike. The callus is an inverted pyramid or square, raised, but no stalk is present. The apically curved column is winged or widened at mid length, with three ridges or keels below the stigma. PLATE 96.

COMMENT *Kefersteinia taurina* is very similar to *K. taggesellii*, with the same color combinations; however, *K. taurina* has horns on the lateral basal edges of the callus though the length of these horns is variable, has a longer isthmus to the lip, has a broader lip, has sepals that do not reflex, and has no wings on the column, compared to *K. taggesellii*, which is more densely spotted centrally on the lip and has definite wings on the column at mid length. These two species could also be considered synonymous as there seem to be intermediates which I have photographed. I have used the presence and absence of wings to tell them apart, but suspect this is just a variable trait in a group of specimens that are very similar in appearance. Many photographs that were considered, labeled *K. graminea* most commonly, but other names also, have turned out to be most consistent with *K taurina* or *K. taggesellii*. *Kefersteinia graminea* has a callus with apical points and its callus is closer to the lip base, whereas *K. taurina* and *K. taggesellii* have a callus that is so elevated it appears to be on a stalk.

MEASUREMENTS Leaves 20 cm long, 1.5 cm wide; inflorescence 6 cm long with one bract; sepals and petals 2 cm long, 0.6 cm wide; lip 1.5 cm long and wide; column 1.3 cm long.

ETYMOLOGY Named for Peter Taggesell.

DISTRIBUTION AND HABITAT Known from Colombia.

FLOWERING TIME Recorded in August.

Kefersteinia taurina (Reichenbach f.)

Reichenbach f. ex Schlechter, *Repert. Spec. Nov. Regni Veg. Beih.* 7: 267. 1920. Basionym: *Zygopetalum taurinum* Reichenbach f., *Linnaea* 41: 5. 1877. Type: Colombia (holotype: W).

DESCRIPTION The flowers are held on upright inflorescences. The sepals and petals are green-white covered with flecks of red, the lip and callus are white with red spots that become more densely concentrated at midline. The obscurely three-lobed lip has an isthmus that broadens gradually to lip mid length where the lip reflexes abruptly. The apical lobe is crispate and has a small apical notch. The callus has a very high base below the lamella of the callus, the flat upper plate of the callus is four-lobed, the posterior lobes are narrowly triangular, and the anterior lobes are semiovate. The column is slightly curved on the dorsum, with three longitudinal keels on the ventral side mid length of the column below the stigma, the lateral keels are less prominent and extend slightly more basal than the medial keel. The column has a prominent but short foot. PLATE 97.

COMMENT The drawing of *Kefersteinia taurina* in *Icones Plantarum Tropicarum* (plate 441) shows thin hornlike appendages on the basal portion of the callus plate. Flowers I have seen in Colombia and Ecuador lack these appendages but possess blunt oblong lobes on the base of the callus plate. In other aspects they are the same species. Pubescence is not mentioned in the original description of this species, but a plant that matches that description was seen and photographed at Ecuagenera in Ecuador; this plant clearly shows short hairs on the ventral column surface. I have also seen this faint pubescence in other photographs of this species.

Kefersteinia taurina is very similar to *K. taggesellii*, however *K. taurina* has horns on the lateral basal edges of the callus, a longer isthmus to the lip, a broader lip, sepals that do not reflex, and no wings on the column; *K. taggesellii* is more densely spotted centrally on the lip and has wings on the column at mid length. *Kefersteinia graminea* and *K. taurina* are also very similar in color pattern, but the callus of *K. graminea* starts at the base of the claw of the lip and is a wavy structure with apical points (like *K. tolimensis*), whereas the callus of *K. taurina* has a base which could almost be called stalklike with a flat plate structure on top. *Kefersteinia graminea* lacks the narrow isthmus after the callus of *K. taurina*. Four (or five) species have a similar curved dorsal column: *K. graminea* (and *K. stapelioides* if you think it is its own species), *K. lehmannii*, *K. taggesellii*, and *K. taurina*.

MEASUREMENTS Leaves 21 cm long, 1 cm wide; inflorescence 5–6 cm long, with two bracts; dorsal sepal 2.2 cm long, 0.7 cm wide; lateral sepals 2.1 cm long, 0.6 cm wide; petals 2.3 cm long, 0.6 cm wide; lip 1.8 cm long, 1.7 cm wide; column 1.4 cm long.

ETYMOLOGY Latin *taurus*, bull, referring to the high callus lobes that look like horns.

DISTRIBUTION AND HABITAT Known from Colombia and western Ecuador, at 1800–2100 m in elevation, in extremely wet montane cloud forest.

FLOWERING TIME Most of the year.

Kefersteinia tinschertiana Pupulin, *Harvard Pap. Bot.* 8(2): 166. 2004. Type: Guatemala, Suchitepéquez, Finca Santa Adelaida, Santa Bárbara, 900 m, October 1982, *M. Dix, M. A. Dix & D. Montúfar 6630* (holotype: MO; isotype: UVAL).

DESCRIPTION The flowers are borne on pendent inflorescences. The sepals and petals are white sometimes with maroon dots, the lip is translucent white with light purple to maroon spots, and the callus is white with faint maroon dots at the base. The flower is bell-shaped to fairly open, the dorsal sepal is concave at its apex, and the lateral sepals are concave. The lip has a very short claw, is concave on the basal half abruptly deflexing at mid length, the apex of the lip is deeply bilobed, and the apical margin is subcrenulate. The bilobed callus is basal, each lobe containing a flat suberect plate with a sulcus between the lobes. The column has inconspicuous auricles at the base and no wings, only a widening lateral to the stigma that continues basally to midcolumn length. The ventral surface of the column has multiple transverse low keels (keels that go across the column horizontally, not running the length of the column) and a median short rounded longitudinal ventral keel. PLATE 98.

MEASUREMENTS Leaves 12 cm long, 1.6 cm wide; inflorescence 2.5 cm long, with two bracts; dorsal sepal 1.2 cm long, 0.5 cm wide; lateral sepals 1.3 cm long, 0.6 cm wide; petals 1.1 cm long, 0.5 cm wide; lip 1.3 cm long, 1.4 cm wide; column 0.9 cm long, 0.4 cm wide.

ETYMOLOGY Named in honor of Otto Tinschert for his outstanding contribution to the study of the orchids of Guatemala.

DISTRIBUTION AND HABITAT Known only from Guatemala, on the Pacific slopes of the Sierra Madres, at 900–1000 m in elevation, in the shade of premontane forest.

FLOWERING TIME October.

Kefersteinia tolimensis Schlechter, *Repert. Spec. Nov. Regni Veg. Beih.* 7: 161. 1920. Type: Colombia, La Plata Vieja, 1500 m, November 1882, *F. C. Lehmann 2205* (holotype: B, destroyed).

DESCRIPTION The flowers are borne on pendent inflorescences. The sepals and petals are green to brown-yellow with brown-red spots, the lip is white-green with brown-red to almost black spots which variably either remain distinct or converge together over the entire surface. The diamond-shaped almost ovate lip is obscurely three-lobed, clawed, and reflexed at mid length, the lip margins have short lacerations and are strongly crispate. The callus is at the base of the lip apical to the small claw, is quadrate to oblong, and is bilobed with variable teeth apically and laterally. The callus typically has four points. The long straight column has no to minimal wings and no to a minimal ventral keel. PLATE 99.

COMMENT *Kefersteinia tolimensis* is the most common species in the genus at shows and it seems almost every species collector has one of these. The pattern of the spots is variable. I have seen plants that are devoid of spots except for a few about the callus and others that are densely spotted. Perhaps the darkest clone I have seen is one that Francisco Villegas grows in Medellín; it is almost black.

I have seen plants of *Kefersteinia tolimensis* labeled as *K. graminea*; they are similar in appearance, especially because they belong to a group of *Kefersteinia* species with flowers of papery substance. *Kefersteinia graminea* has a callus that is wider than long, becoming wider apically, with four points; *K. tolimensis* has a squarer callus with lateral points at its four corners. The straight column of *K. tolimensis* is plain having no wings or ventral keel or tooth, whereas the dorsally curved column of *K. graminea* has minimal mid column wings and a ventral keel that becomes a tooth at mid column length.

The Illustrated Encyclopedia of Orchids (Pridgeon 1992) has a specimen labeled *Kefersteinia tolimensis* with long white trichomes that I believe is either *K. pellita* or *K. villosa*, but it is not *K. tolimensis* based on the trichomes and the obvious wings on the column.

MEASUREMENTS Leaves 20–25 cm long, 2.5–2.7 cm wide; inflorescence 10 cm long with one bract; sepals 2.3 cm long, 1 cm wide; petals 2 cm long, 1.2 cm wide; lip 1.8 cm long, 1.9 cm wide; column 1.1 cm long.

ETYMOLOGY Named for the type locality of Tolima, Colombia.

DISTRIBUTION AND HABITAT Known from Colombia, Ecuador, and Venezuela, at 1500–2100 m in elevation, in warm moist habitat.

FLOWERING TIME Several times a year.

Kefersteinia trullata Dressler, *Orquideología* 18: 221. 1993. Type: Colombia, Chocó, El Embarcadero, 1000 m, collected by Julio César Miranda, 1988, flowered in cultivation at Colomborquídeas, 6 July 1989, *R. Escobar 4023* (holotype: JAUM; isotype: MO).

SYNONYM
Senghasia trullata (Dressler) Szlachetko, *J. Orchideenfr.* 10: 336. 2003.

DESCRIPTION The flowers are borne on semierect to pendent inflorescences. The dorsal sepal is yellow-green with small red markings, the lateral sepals are light yellow, each with a large wine-red blotch, the petals are very dark red, the lip is red with very dark lateral margins, and the column is green and stained red. The lip is without a claw and is trullate. The lip base is concave, apically and laterally becoming flat but not deflexing, and the lip apex is attenuate. The callus is stalked and two-lobed. The column is basally flattened, the ventral keel is rudimentary or lacking, and the plate below the stigma is winglike on each side. PLATE 100.

MEASUREMENTS Leaves 13.0–16.5 cm long, 1.2–1.7 cm wide; inflorescence 3.0–4.5 cm long, with two bracts; dorsal sepal 1.8 cm long, 0.6 cm wide; lateral sepals 2.1 cm long, 0.7 cm wide; petals 0.9 cm long, 0.3 cm wide; lip 0.8 cm long, 0.6 cm wide; column 0.7 cm long.

ETYMOLOGY Latin *trullatus*, angular ovate, the shape of a brick layer's trowel, referring to the outline of the lip.

DISTRIBUTION AND HABITAT Known only to Colombia and maybe Ecuador, at 1000–1500 m in elevation.

FLOWERING TIME January to March.

Kefersteinia urabaensis Aguirre & P. A. Harding, *Orquideología* 26(1). 2008. Type: Colombia, Antioquia, Urabá, collected by Dr. Rogelio Londoño (holotype: JAUM).

DESCRIPTION The inflorescences are pendent with single nonresupinate flowers. The sepals and petals are green-white with a few maroon spots at the base, the lip is white with diffuse to heavily coalescing red-maroon spots over its surface sparing the lip margins. The column is green with maroon spots at the base. The anther cap is yellow-white. The dorsal sepal is oblong and blunt at the apex. The lateral sepals are oblong with blunt apexes. The petals are ovate and blunt. The lip is unlobed with a very small claw, transversely sharply folded at mid length to form a skirt, denticulate at its apical margins, and notched at its apex. The lip callus is wedge-shaped, with lateral margins that appear notched at mid length. The callus apexes diverge laterally and extend apically to a quarter of the lip length. The column is straight, has a minimal raised ventral keel that extends to the column foot, and has blunt wings at the column mid length that are held in the same plane as the column. There is a second widening, or auricle, at the column base lateral to the column foot, and the angle of the column foot to the column is 90 degrees. The column foot has no apical tooth. The column is pubescent on the ventral surface in the area of the column wings and medially. PLATE 101.

COMMENT *Kefersteinia klabochii* is very similar to *K. urabaensis* but the base of the column is different and the shape of the wings is different, though both species hold their wings out laterally, not angling ventrally. The two also differ in their callus: that of *K. klabochii* is scalloped or multinotched, but the callus of *K. urabaensis* is notched only once. The skirt fold of *K. klabochii* also forms more of a circle than *K. urabaensis*, and the color pattern of the two species also differs slightly. Still, all considered, I think *K. urabaensis* could be a synonym or form of *K. klabochii* as they are so similar.

Kefersteinia urabaensis is very similar to *K. escalerensis* from Peru. Both species have a short column with a ventral keel that continues into the column foot; in both the angle between the column foot and the column is 90 degrees, and both have basal auricles on the column. *Kefersteinia urabaensis* differs most notably by having a flower that is always nonresupinate and a callus that lacks the nipplelike apical ends of *K. escalerensis*. It also differs in that its wings do not curl while those of *K. escalerensis* curl down and in. *Kefersteinia urabaensis* lacks the ventral tooth found on the column foot of *K. escalerensis*.

Kefersteinia urabaensis differs from *K. laminata* and *K. gemma* by the smaller ratio of callus length to lip length. The calluses of *K. laminata* and *K. gemma* are nearly one-third to one-half the lip length and extend apically almost to the deflection of the lip, whereas the callus of *K. urabaensis* extends apically only about a quarter of the lip length.

Kefersteinia urabaensis differs from *K. sanguinolenta* by being nonresupinate and by having prominent earlike widening at the base of the column. *Kefersteinia sanguinolenta* has a prominent ventral keel that extends basally only about half the column length, whereas the ventral keel of the column of *K. urabaensis* extends through to the column foot. The wings of *K. sanguinolenta* angle downward toward the callus as opposed to the wings of *K. urabaensis*, which are held out, creating a flat ventral column surface excepting the slightly elevated ventral tooth.

MEASUREMENTS Leaves 15 cm long, 1.5 cm wide; inflorescence 4 cm long; dorsal sepal 1.5 cm long, 0.5 cm wide; lateral sepals 1.7 cm long, 0.5 cm wide; petals 1.5 cm long, 0.2 cm wide; lip 1 cm long, 1.3 cm wide.

ETYMOLOGY Named for the region where the type plant was collected.

DISTRIBUTION AND HABITAT Known only from Colombia; its original habitat is unknown.

FLOWERING TIME Sporadically throughout the year.

Kefersteinia vasquezii Dodson, Icon. Pl. Trop., ser. 2, 4: plate 345. 1989. Type: Bolivia, La Paz, Inquisive, near Circuata, flowered in cultivation 22 December 1983, *R. Vásquez, C. & J. Luer & Besse 821* (holotype: MO).

DESCRIPTION The flowers are held on upright inflorescences and are green-white with a large dark purple spot in the middle of the lip. The lip is suborbicular, abruptly reflexed at mid length, with a thickened rim at the point of reflection, the lip margin is serrate, and the lip is notched at the apex. The callus lobes are narrow and diverge from the base to form two horns. The column has a minimal ventral keel below the stigma and long wings or widenings on each side that make the ventral surface contained appear square.

COMMENT The coloration of *Kefersteinia vasquezii* is the same as that of *K. pulchella*. The velvety glandular texture of the lip in *K. pulchella* is not discussed in the description of the type of *K. vasquezii*. The drawing of the type of the species of each varies a little in minor ways that could be accounted by artistic interpretation. I have left these two spectacularly colored and dramatic species separate, but feel the reality is that *K. vasquezii* is a synonym or a subspecies of *K. pulchella*. The type drawings show a notch at the apex of the lip of *K. vasquezii*, a broader, more rhomboid ventral surface of the column, and sepals and petals that recurve back at their apexes. If there is a notch on the lip of *K. pulchella*, it is subtle, the flat plate on the ventral surface of the column is more pear-shaped with the widest portion being at the plates base, and the sepals and petals are not recurved.

MEASUREMENTS Leaves 10 cm long, 1.5 cm wide; inflorescence 4.0–4.5 cm long, with one or two bracts; dorsal sepal and petals 2 cm long, 0.8 cm wide; lateral sepals 2.3 cm long, 0.8 cm wide; lip 2 cm long, 2.4 cm wide; column 1.5 cm long.

ETYMOLOGY Named for Roberto Vásquez, who discovered the species and drew the type illustration.

DISTRIBUTION AND HABITAT Known only from Bolivia and Peru, elevation not recorded, in montane wet forest.

FLOWERING TIME December to January.

Kefersteinia villenae D. E. Bennett & Christenson, *Brittonia* 46: 241. 1994. Type: Peru, San Martín, Moyobamba, near Pacaysapa, 1100 m, 10 September 1992, *R. Villena & B. Collantes ex Bennett 5699* (holotype: NY; isotype: MOL).

DESCRIPTION The flowers are borne on pendent inflorescences. The sepals are very pale green-yellow with purple-red spots which sometimes coalesce, the lip is pale tan to off-white with dark red spots, the column is pale yellow-green with ventral brown-red spots, and the anther is pale yellow or white. The sepals and petals are long and thin. The acuminate petals are recurved slightly at their apex. The lip is obovate, recurved sharply at midpoint, with erose, denticulate apical margins. The callus is subrhombic, arcuate, stalkless, and bilobed, each lobe with apical and lateral points. The column is winged below the middle, and the lightly pubescent wings are directed basally and slightly downward. The column has a ventral keel that is highest below the stigma diminishing toward the convex base of the column foot. PLATE 102.

COMMENT *Kefersteinia villenae* resembles the densely spotted pubescent species, but lacks the long trichomes, and the longest hairs one can see are no more than just fuzz on the ventral surface of the column. Particularly, the shape of the sepals and petals and the callus of *K. villenae* are very similar to those of *K. escobariana*, but *K. villenae* lacks the prominent long white trichomes or hairs of *K. escobariana*. Plate 102 in the present volume seems to fit this species and was submitted by Carlos Hajek of Peru, but I can make no definitive identification.

MEASUREMENTS Leaves 8.0–11.2 cm long, 1.0–1.2 cm wide; inflorescence 4 cm long, with two bracts; sepals 1.5–1.7 cm long, 0.5 cm wide; petals 1.5 cm long, 0.6 cm wide; lip 1.2 cm long, 0.9 cm wide; column 0.8 cm long, 0.4 cm wide.

ETYMOLOGY Named for Renato Villena, who collected the species.

DISTRIBUTION AND HABITAT Known only from Peru, at 1100 m in elevation, in montane rain forest in deep shade.

FLOWERING TIME Throughout the year as growths mature.

Kefersteinia villosa D. E. Bennett & Christenson, *Lindleyana* 13: 51. 1998. Type: Peru, Junín, Chanchamayo, Arcopunco, above Quebrada Seca, 2500 m, collected by O. del Castillo 20 June 1992, flowered in cultivation 28 August 1993, *Bennett 5609* (holotype: NY; isotype: MOL).

DESCRIPTION The flowers are held on upright inflorescences. The flower is spotted externally in the bud, the sepals and petals are green with purple to red-purple blotches and spots, the lip is pale green with dark maroon blotches and spots, the callus is pale rose, the column is pale green with small dark purple blotches, the anther is pale lime green with a purple blotch on each side, and the pollinia are pale yellow. The ovate lip has a short claw, is transversely deflexed at the middle forming a skirt, and the disc of the lip has hairs over most of its surface. The callus has two rounded, subauriculate lobules which are bluntly pointed apically, and the callus surface is hirsute villose. The column is arcuate slightly, fleshy, stout, and pubescent, with rounded wings parallel to the ventral keel elevation; the keel starts at the stigma and extends basally to mid column length.

COMMENT When *Kefersteinia villosa* was described, the illustration of *K. pellita* (which is incorrectly labeled and is more correctly *K. pseudopellita*) in *Icones Plantarum Tropicarum* (1989, series 2, plate 86) was used as a reference (Christenson, pers. comm.), causing them to create the new species *K. villosa* instead of considering *K. villosa* a potential specimen of *K. pellita*. I have not seen a live plant of either *K. villosa* or *K. pellita*, or a photograph that I am certain is either species. I don't know whether the angle of the callus lateral edges or the auricles on the callus are significantly different in these two species compared to *K. pseudopellita*, which I know from live material and multiple photographs. All three species have the column foot in almost the same plane as the column; they have a midline ventral tooth on the column that does not descend basally, wings that extend laterally from the mid column rather than angling

down or ventrally, along with sharing the same villose lip and color pattern. *Kefersteinia pellita* would have name priority if *K. villosa* is determined to be a synonym.

Kefersteinia villosa is very similar to *K. aurorae* and *K. pastorellii* with a few subtle differences. *Kefersteinia villosa* has the column foot in the same plane as the main part of the column. *Kefersteinia aurorae* has a 150-degree angle between the column foot and the column, and *K. pastorellii* has a 90-degree angle between the column foot and the column. The columns of *K. villosa* and *K. pastorellii* are more oblong (being twice as long as wide) compared to the column of *K. aurorae*, which is more square (ratio of length to width is 3:2).

MEASUREMENTS Leaves 20 cm long, 1.4 cm wide; inflorescence 5 cm long with one bract; dorsal sepal 1.2 cm long, 0.5 cm wide; lateral sepals and petals 1.5 cm long, 0.6 cm wide; lip 1 cm long and wide; column 0.8 cm long, 0.4 cm wide.

ETYMOLOGY Latin *villosus*, long slender hairs, describing the ornamentation of the lip surface.

DISTRIBUTION AND HABITAT Known from Peru, at 2500 m in elevation, in wet montane cloud forest.

FLOWERING TIME Late August to October.

Kefersteinia vollesii Jenny, *Orchidee (Hamburg)* 36(5): 185. 1985. Type: Colombia, Mocoa, 950 m, collected by T. Neudecker, 1985, *R. Jenny 51* (holotype: G).

SYNONYM
Senghasia vollesii (Jenny) Szlachetko, *J. Orchideenfr.* 10(4): 336. 2003.

DESCRIPTION The flowers are held on semiupright inflorescences. The sepals and petals are yellow-green with dark red-brown spots, the lip is green with dark red spots becoming confluent medially and sparser at the margins. The flower is somewhat bell-shaped, with sepals and petals that do not spread fully. The dorsal sepal covers the column from above. The lateral sepals are longer than the dorsal sepal and petals. The lip is denticulate, crispate, and ovate when flattened, but in its natural form the lip is compressed (pinched) midway. The callus has two lobules at the very base of the lip, and the basal lateral points of the callus extend laterally to the lip edges. PLATE 103.

COMMENT Günther Gerlach and Tilman Neudecker (1994) have this species as a synonym of *Kefersteinia pusilla*.

MEASUREMENTS Leaves 15 cm long, 3 cm wide; inflorescence 4 cm long with one bract; sepals 1.2–1.3 cm long, 0.4 cm wide; petals 1.3 cm long, 0.4 cm wide; lip 1.2 cm long, 0.7 cm wide; column 1.2 cm long.

ETYMOLOGY Named for Hans Volles, a German orchid enthusiast in Colombia who participated in the collection of the original plant.

DISTRIBUTION AND HABITAT Known from Colombia, at 950 m in elevation.

FLOWERING TIME Not recorded.

Kefersteinia wercklei Schlechter, *Repert. Spec. Nov. Regni Veg. Beih.* 19: 53. 1923. Type: Costa Rica, La Palma, June 1921, *C. Wercklé 120* (holotype: B, destroyed; lectotype: AMES).

SYNONYMS
Chondrorhyncha wercklei (Schlechter) C. Schweinfurth, *Amer. Orchid Soc. Bull.* 12: 386. 1944.
Kefersteinia umbonata Reichenbach f. ex Senghas & Gerlach, *in syn., Orchideen* (Schlechter) 1/B(26): 1641. 1992.

Senghasia wercklei (Schlechter) Szlachetko, *J. Orchideenfr.* 10: 337. 2003.
Zygopetalum umbonatum Reichenbach f., Hort Och Nr 49841 not published.

DESCRIPTION The flowers are borne on pendent inflorescences and are pale yellow-green with red or pink spots on the sepals and petals and a pink suffusion or red-pink on the white lip. The sepals and petals are five-veined. The petals are slightly shorter than sepals. The lip is oblong, fiddle-shaped, the lateral edges curving up, the fan-shaped apical end of the lip mildly reflexing down, and the apical lip margin is crenulate. The stalked callus is bilobed, with each lobe having its own sulcus. The column is semicircular, the apex is dilated, and the clinandrium is mildly elevated on the dorsum. The column is widest at its stigma without wings, and the ventral keel starts at the stigma, extends through the nontoothed, four-pointed, and square ventral plate and continues to the base of the column. PLATE 104.

COMMENT Two photographs of this species taken by Franco Pupulin are posted at www.neotrop.org/gallery.

MEASUREMENTS Leaves 10–17 cm long, 1.2–1.5 cm wide; inflorescence 3–4 cm long, with three or four bracts; sepals and petals 1.0–1.2 cm long, 0.6 cm wide; lip 0.9 cm long, 0.5 cm wide; column 0.7–0.8 cm long.

ETYMOLOGY Named for C. Wercklé, who collected the original plant.

DISTRIBUTION AND HABITAT Recorded from Costa Rica, on the Atlantic side near La Palma.

FLOWERING TIME Recorded in June.

• •

List of *Kefersteinia* Species with Showy Skirts

In this section I have tried to separate those species with lips that are folded to form a skirt, segregating them by the color of the flowers or other characteristics that would make them tricky to separate from other species based on a quick look. The hope is that you will look at this list along with the key at the beginning of the chapter to help narrow your choices and then read the species description.

Species with Flowers Spotted Brown or Maroon

Showy skirts of flowers covered with dark red, dark brown, or maroon spotting, the base color sometimes barely showing through the spots on the sepals, petals, and lips.

Kefersteinia aurorae, lip with long hairs, column with small lateral wings, ventral keel extends to column foot, column foot at 150-degree angle to column, dorsal sepal broad.

Kefersteinia bismarckii, callus widens then narrows, apical teeth acute, column foot at 150-degree angle to column, lip pubescent, column not pubescent.

Kefersteinia escobariana, wings low on the column, low ventral tooth, callus fused to one apical point.

Kefersteinia graminea, column curved apically, callus wider than long, wings minimal, ventral column keel prominent.

Kefersteinia koechliniorum, lip reflexes severely on itself.

Kefersteinia pastorellii, column short, flower bell-shaped, lip margin white, column foot at 90-degree angle to column, column and lip pubescent.

Kefersteinia pellita, lip and ventral column with long hairs, ventral keel not extending to column base, sepals densely marked maroon or brown, dorsal sepal long and thin, callus with lateral auricles.

Kefersteinia pseudopellita, lip and ventral column with long hairs, ventral keel not extending to column base, sepals densely marked maroon or brown, callus without lateral auricles.

Kefersteinia tolimensis, column wingless, ventral keel extends to base, long straight column juts out of flower center.

Kefersteinia villenae, column with large ventral tooth, apical to the wings, wings low on the column.

Kefersteinia villosa, lips with hairs, wings at mid column length, callus pale rose, ventral keel not extending to column base, column foot and column main in same plane, callus with basal lateral lobules and acute apexes.

Species with White Flowers

Flowers pure white, not light yellow or green, with or without spots. Includes some slightly off-white flowers.

Kefersteinia alba, pink spots at base of lip, rim of skirt deflection forms a third of a circle, callus widens then narrows, wings and ventral keel present.

Kefersteinia bengasahra, dark purple spots on lip, pubescent, ventral keel.

Kefersteinia candida, all white, rim of skirt deflection forms half a circle, callus with multiple blunt angles on lateral edge.

Kefersteinia forcipata, solid brick-red coloration lateral to the callus.

Kefersteinia klabochii, pink or purple spots, wings on column do not curl ventrally.

Kefersteinia lactea, brown or red spots at base of lip, rim of skirt deflection forms half a circle, callus widens apically, column with wings.

Kefersteinia lafontainei, all white with yellow wings on column, rim of skirt deflection forms over half a circle, callus has four acute points.

Kefersteinia microcharis, pink spots or dots on lip, blotches and streaks medially (rim of skirt deflection forms half a circle), callus widens toward apex, ventral tooth on column.

Kefersteinia richardhegerlii, maroon spots diffuse on sepals and petals, lip margin free of spots.

Kefersteinia tinschertiana, small purple spots, more on the lip, column wingless, callus apexes rounded.

Kefersteinia urabaensis, flower nonresupinate, white with pink spots, wings on column don't extend ventrally, ventral keel.

Species with Off-white or Green-White Flowers

Some flowers are heavily marked on the tepals.

Kefersteinia escalerensis, sepals and petals yellow-white; few rose-purple spots on lip, sepals, and petals; callus apical tips nipplelike; wings at mid length of column; ventral keel extends to base with apical tooth and auricles at base of column.

Kefersteinia graminea, lip white or very pale green with red spots, callus wider than long, no or minimal wings.

Kefersteinia heideri, red color in rays or lines on lip.

Kefersteinia pulchella, green-white with one large maroon spot on lip.

Kefersteinia sanguinolenta, wings at column mid length, column ventral keel extends to mid column length only.

Kefersteinia stapelioides, callus longer and thinner than *K. graminea*.

Kefersteinia taggesellii, sepals and petals green with red spots, lip white base centrally completely covered with dark maroon spots.

Kefersteinia taurina, sepals and petals green with red spots, lip white base covered with red spots, thicker spotting centrally.

Kefersteinia vasquezii, flower green-white with one large maroon spot on lip.

Kefersteinia villenae, lip bell-shaped, column ventral keel extends to base.

Species in the Gemma Complex

I do not suppose to state that this a natural complex. These flowers have lips that, rather than extend horizontally from the lip base, extend upward, and at mid length the lip reflexes back more than 270 degrees, almost folding on itself. These small flowers have a callus that extends to nearly the transverse fold of the lip.

Kefersteinia andreettae, flowers yellow sometimes with small red spots.

Kefersteinia atropurpurea, flowers green with transverse dark purple markings.

Kefersteinia delcastilloi, spots purple, lip white with few medial brown spots, callus with medial hump between lobes.

Kefersteinia gemma, flowers green or yellow-green with pink spots, finely pubescent, callus with four point, lateral points acute.

Kefersteinia laminata, sepals and petals yellow, lip white with numerous purple spots, callus with four points, lateral points blunt.

Kefersteinia oscarii, flowers green or yellow-green with more pink spots on lip than *K. gemma*, callus with two points.

Kefersteinia punctatissima, callus with six points—two lateral at base and four at apex.

Species by Country of Origin

Bolivia
Kefersteinia heideri
Kefersteinia mystacina
Kefersteinia pulchella
Kefersteinia richardhegerlii
Kefersteinia ricii
Kefersteinia sanguinolenta
Kefersteinia vasquezii

Brazil
Kefersteinia mystacina

Colombia
Kefersteinia chocoensis
Kefersteinia elegans
Kefersteinia forcipata
Kefersteinia gemma
Kefersteinia graminea
Kefersteinia klabochii
Kefersteinia laminata
Kefersteinia lehmannii

Kefersteinia mystacina
Kefersteinia niesseniae
Kefersteinia ocellata
Kefersteinia oscarii
Kefersteinia parvilabris
Kefersteinia perlonga
Kefersteinia sanguinolenta
Kefersteinia taggesellii
Kefersteinia taurina
Kefersteinia tolimensis
Kefersteinia trullata
Kefersteinia urabaensis
Kefersteinia vollesii

Costa Rica
Kefersteinia alba
Kefersteinia costaricensis
Kefersteinia endresii
Kefersteinia excentrica
Kefersteinia lactea
Kefersteinia microcharis

Kefersteinia orbicularis
Kefersteinia parvilabris
Kefersteinia retanae
Kefersteinia wercklei

Ecuador
Kefersteinia andreettae
Kefersteinia escobariana
Kefersteinia expansa
Kefersteinia forcipata
Kefersteinia gemma
Kefersteinia graminea
Kefersteinia guacamayoana
Kefersteinia hirtzii
Kefersteinia laminata
Kefersteinia lehmannii
Kefersteinia lindneri
Kefersteinia lojae
Kefersteinia minutiflora
Kefersteinia mystacina
Kefersteinia ocellata

Kefersteinia pellita
Kefersteinia pseudopellita
Kefersteinia sanguinolenta
Kefersteinia stevensonii
Kefersteinia taurina
Kefersteinia tolimensis
Kefersteinia villosa

French Guiana
Kefersteinia lafontainei

Guatemala
Kefersteinia lactea
Kefersteinia tinschertiana

Mexico
Kefersteinia lactea

Nicaragua
Kefersteinia costaricensis
Kefersteinia lactea

Panama
Kefersteinia alata
Kefersteinia alba
Kefersteinia angustifolia
Kefersteinia auriculata
Kefersteinia costaricensis
Kefersteinia elegans
Kefersteinia excentrica
Kefersteinia lactea
Kefersteinia maculosa
Kefersteinia mystacina
Kefersteinia orbicularis
Kefersteinia parvilabris

Peru
Kefersteinia andreettae
Kefersteinia atropurpurea
Kefersteinia aurorae
Kefersteinia bengasahra
Kefersteinia benvenathar
Kefersteinia bertoldii

Kefersteinia bismarckii
Kefersteinia candida
Kefersteinia delcastilloi
Kefersteinia escalerensis
Kefersteinia jarae
Kefersteinia koechliniorum
Kefersteinia licethyae
Kefersteinia mystacina
Kefersteinia pastorellii
Kefersteinia pellita
Kefersteinia punctatissima
Kefersteinia pusilla
Kefersteinia sanguinolenta
Kefersteinia villenae
Kefersteinia villosa

Venezuela
Kefersteinia graminea
Kefersteinia sanguinolenta
Kefersteinia stapelioides
Kefersteinia tolimensis

Pescatorea Reichenbach f., *Bot. Zeit. (Berlin)* 10: 667. 1852.

The first *Pescatorea* species, *P. cerina*, was initially described by John Lindley in 1852 within the genus *Huntleya* as *H. cerina*. Later that same year Heinrich Gustav Reichenbach moved *H. cerina* to his new genus *Pescatorea*, dedicated to M. Pescatore (1793–1855). The original description was first published with the spelling *Pescatoria*. Reichenbach later changed the spelling to *Pescatorea*, without comment on the previous spelling, which is orthographically more correct. None of the subsequent species described nor authors discussing the genus used the spelling *Pescatoria* until recently. Robert Dressler makes an excellent argument for using the spelling *Pescatorea*, as the first spelling is considered a typographical or orthographical error (Whitten et al. 2005). The International Code of Botanical Nomenclature recommends that the form be *Pescatorea*, adding an *-a* to a person's proper name. Please note, however, that the Kew World Checklist of Monocotyledons lists the name as *Pescatoria*, and International Plant Name Index and w³TROPICOS list both spellings.

The genus *Pescatorea* is comprised of fairly large plants that grow as tufted epiphytic herbs without pseudobulbs. The leaves are plicate, contracted below into conduplicate petioles, distichously arranged in the form of an open fan. The inflorescence is produced from the axis of the leaf sheath, and the flower is presented erect to laxly pendant, generally below the overhanging foliage. The sepals are more or less fleshy, subequal (the lateral sepals are somewhat larger than the dorsal sepal), and concave; the dorsal sepal is erect and free; the lateral sepals are connate at the base and obliquely inserted onto the column foot. The petals are subequal but smaller than the lateral sepals. The one- to three-lobed lip is very fleshy. The lip is contracted at the base into a conspicuous claw that is contiguous with the column foot. The base of the lip forms a deep concavity below the column; the concavity is surrounded from below by an erect multikeeled callus. The apical lobes of the lip are convex or ventricose and never hooded by the column, and the lateral margins of the midlobes of the lip are in most species recurved, some to an extreme. The column is stout, hirsute or glabrous, and has wings forming a concavity on the ventral surface of the column, and the column foot is short. The anther is terminal, operculate, incumbent, and one-celled. The four pollinia are flat and waxy, on a trullate viscidium with a minimal stipe.

Later in the year that *Pescatorea* was established as a separate genus by Reichenbach, the genus *Bollea* was also established by Reichenbach. *Bollea* was considered to be similar to *Pescatorea* but with significantly broader columns that cover the callus and basal lip from above. *Bollea*, dedicated to Karl Boll, a German horticulturalist, consists of plants that are vegetatively similar to *Pescatorea*, lacking pseudobulbs and having foliage that is distinctly sheathed and appearing like a fan. The leaves are slender and slightly ribbed. The inflorescence is one-flowered, emerging from the axis of the leaf sheath, and is semierect but remaining under the foliage in height. The flower lip has a large fleshy callus with parallel crests, and the callus fits within the wings of the overhead column. There are four flat pollinia on a short stipe, which is attached to a viscidium that is flat and heart-shaped. Except for the callus fitting within the wings of the overhead column, this is the same description as for *Pescatorea sensu stricto*.

Molecular sampling of *Pescatorea* and *Bollea* shows that the species fall in an interspersed manor on the cladogram, making it impossible to separate the two genera. Whitten et al. (2005) combined the species of *Bollea* and *Pescatorea*, selecting *Pescatorea* as the accepted name (and spelling). The genus ranges from Costa Rica to Peru.

Pescatorea is known for being variable in coloration and shape within the individual species; colors given in the key and text are the "typical color." Some of the color "variation" is undoubtedly also due to misidentification and/or natural hybrids, but there is a large range of color variation also. Even though the shape of the flowers varies within the species, certain characteristics make it possible to distinguish one species from another. In the wild, natural hybrids of different species of *Pescatorea* (and the old genus *Bollea* with *Pescatorea*) exist; some of these are mentioned in the text at the end of the chapter. Please take all colors mentioned here as what is most common not what is a rule.

Etymology
Named for M. Pescatore (1793–1855), a French orchid collector and horticulturalist.

List of the Species of *Pescatorea*

*Pescatorea cerina** (type species of *Pescatorea*)

Pescatorea cochlearis

Pescatorea coelestis

*Pescatorea coronaria**

Pescatorea dayana

Pescatorea ecuadorana*   Pescatorea lawrenceana*
Pescatorea hemixantha   Pescatorea lehmannii*
Pescatorea hirtzii   Pescatorea pulvinaris*
Pescatorea klabochorum*   Pescatorea violacea
Pescatorea lalindei   Pescatorea wallisii
Pescatorea lamellosa*

* Molecular sampling confirms placement of this species within the genus *Pescatorea*.

Some "species" are hybrids or most likely synonyms or aberrant forms of species, but as they are often listed as species in the literature and some are even listed as accepted names on the Kew World Checklist of Monocotyledons, these species listed below are covered in the text at the end of the chapter.

Pescatorea backhousiana   Pescatorea pallens
Pescatorea bella   Pescatorea rueckeriana
Pescatorea dormaniana   Pescatorea russeliana
Pescatorea gairiana   Pescatorea triumphans
Pescatorea kalbreyeriana   Pescatorea whitei

Key to the Species of *Pescatorea*

   1a. Column with wings that cover the lateral sides of the lip callus, these are the "old" species of *Bollea*. . . . . go to 2

   1b. Column does not cover the lateral sides of the lip callus, these are the "old" species of *Pescatorea* . . . . go to 9

   2a. Clinandrium completely hidden beneath apex of column, so that the anther cap is not seen from the frontal view, column covers callus from above . . . . go to 3

   2b. Clinandrium not completely hidden beneath apex of column, so that the anther cap is visible from the frontal view, callus visible from above . . . . go to 5

   3a. Base of lip before callus with dense warty asperities and hairs; sepals and petals white or yellow, inferior half of lateral sepals same color as superior half; callus of 16 keels. . . . . *P. hemixantha*

   3b. Base of lip before callus without asperities or hairs; sepals and petals purple or pink, callus of 16 or fewer keels. . . . . go to 4

   4a. Callus central keels form three distinct teeth at their apex; flower dark purple including lip, callus white and with 10 keels . . . .*P. violacea*

   4b. Callus central lamella form a fused ridge at their apex, callus with 14 to 16 keels. . . . . *P. lalindei*

   5a. Auricles present on base of column . . . . go to 6

   5b. Auricles lacking on base of column . . . . go to 7

6a. Column and callus widest of the genus, 1.8 cm or more wide; lip velvet red apical to salmon-pink-colored callus . . . .*P. pulvinaris*

6b. Column and callus 1–1.5 cm wide, sepals and petals white or pale pink, column dark pink, red, or purple. . . . . *P. hirtzii*

7a. Callus longer than wide, lip with a long point, sepals and petals with purple, column not solid pink or purple. . . . .*P. lawrenceana*

7b. Callus as wide as or wider than long . . . . go to 8

8a. Base of column with acute posterior appendages and acute ventral tooth at base of column, callus with 12 keels . . . .*P. ecuadorana*

8b. Base of column with blunt posterior appendages and blunt ventral tooth at base of column, callus with 20 keels . . . .*P. coelestis*

9a. Labellum with hairs, warts, or bristles . . . . go to 10

9b. Labellum without hairs, warts, or bristles . . . . go to 15

10a. Lip with long hairs with or without warts . . . . go to 11

10b. Lip with hard warts and no long hairs . . . . go to 13

11a. Callus keels all separate, not forming a confluent structure in the central callus . . . . *P. coronaria*

11b. Callus with three central keels almost confluent into one structure . . . . go to 12

12a. Lip apex convex . . . .*P. lehmannii*

12b. Lip apex concave . . . . *P. cochlearis*

13a. Lip one-lobed, lip apex white with brown lamellae, 13–15 keels . . . .*P. lamellosa*

13a. Lip three-lobed. . . . . go to 14

14a. Sepals and petals yellow, keels brown-red . . . . *P. cerina*

14b. Sepals and petals white with red apexes, warts tipped with red . . . . *P. klabochorum*

15a. Lip roughly square or oval, as long as wide . . . . *P. dayana*

15b. Lip longer than wide, apex heart-shaped . . . . *P. wallisii*

Pescatorea cerina (Lindley & Paxton) Reichenbach f., *Bot. Zeit. (Berlin)* 10: 667. 1852. Basionym: *Huntleya cerina* Lindley & Paxton, *Flow. Gard.* 3: 62. f. 263. 1852. Type: Panama (holotype: W); type species of *Pescatorea*.

SYNONYMS

Pescatorea cerina var. *guttulata* Reichenbach f., *Ann. Bot Syst.* 6: 608. 1863.

Zygopetalum cerinum (Lindley) Reichenbach f., *Ann. Bot. Syst.* 6: 651. 1863.

Pescatorea costaricensis Schlechter, *Repert. Spec. Nov. Regni Veg. Beih.* 19: 139. 1923. Type: Costa Rica, near Standortsangabe, May 1910, *A. C. Brade 1196* (lectotype: AMES).

DESCRIPTION The inflorescence is laxly pendent. The dorsal sepal and petals are white to cream-yellow, the lateral sepals are cream-yellow or white with a long green-yellow blotch near the base, the lip is a rich yellow with a basal callus marked with keels or lamina of red-brown, the column is white, and the anther is lavender. The sepals are wedge-shaped at the base and dilate abruptly at the apex. The three-lobed lip is oval with lateral lobes that are small auricle-like and somewhat erect. The basal portion of the lip is slightly constricted and rolled under. The apical lip surface is very rough, like course sandpaper, but the warts are the same color as the underlying lip. The callus is fan-shaped, with 13 to 15 keels, and the keels are minutely warty. The column is glabrous. PLATE 105.

MEASUREMENTS Leaves 15–60 cm long, 3–5 cm wide; inflorescence 3.5–10.0 cm long; dorsal sepal 2.5–3.2 cm long, 1.6–1.8 cm wide; lateral sepals 2.5–3.5 cm long, 1.8–2.0 cm wide; petals 2.5–3.0 cm long, 1.5–1.8 cm wide; lip 2–3 cm long, 2.5 cm wide; column 1.3–1.5 cm long.

ETYMOLOGY Latin *cerinus*, dull yellow with a soft mixture of red-brown, referring to the color of the sepals, petals, and lip.

DISTRIBUTION AND HABITAT Known from Colombia to Costa Rica, at 750–1200 m in elevation, in wet highland forests of the Pacific slope, usually in shaded situations.

FLOWERING TIME May and June.

Pescatorea cochlearis Rolfe, *Bull. Misc. Inform. Kew*: 33. 1906. Type: Andes (holotype: K).

DESCRIPTION The inflorescence is laxly pendent. The flower is red with white sepals and petal tips, the lip is white with white hairs and a white margin, the callus is dark red. The three-lobed lip is concave at the apex, with long glandular hairs over the apical lip. The callus is composed of three central keels that combine to form one central keel and lateral keels that remain separate (similar to the calluses of *Pescatorea lehmannii* and *P. dormaniana*). PLATE 106.

COMMENT The original description of this species, with an accompanying picture drawn to confirm the colors, says that the flower has white sepals and petals with red-maroon

apexes, a white lip with crest tubercles, and a maroon column. This specimen was obtained from the collection of Ida Brandt of Zurich. A picture of this species in *Orchid Digest* (1969, 33: 100) has many of the characteristics that would be considered *Pescatorea coronaria* except that the lip is concave instead of convex.

MEASUREMENTS Leaves 13.5–24.3 cm long, 2.0–3.4 cm wide; inflorescence 9 cm long; sepals 3.4 cm long, 2.5 cm wide; petals 3.4 cm long, 1.8 wide; lip 2.7 cm long, 2.3 cm wide; column 2.1 cm long.

ETYMOLOGY Latin *cochleatus*, spoon-shaped, referring to the shape of the lip.

DISTRIBUTION AND HABITAT Known from Colombia and Ecuador, at 1000 m in elevation.

FLOWERING TIME Not recorded.

Pescatorea coelestis (Reichenbach f.) Dressler, *Lankesteriana* 5(2): 95. 2005. Basionym: *Bollea coelestis* Reichenbach f., *Gard. Chron.* 2(5): 756. 1876. Type: Colombia (holotype: W).

SYNONYM
Zygopetalum coeleste (Reichenbach f.) Reichenbach f., *Linnaea* 41: 5. 1876.

DESCRIPTION The inflorescence is laxly pendent to upright. The sepals and petals are purple to dark purple, sometimes blue to blue-purple, with white apexes, the lip is red-purple to purple, the callus is yellow, and the column is purple with a lighter underside, often yellow. The sepals and petals are elliptical to spatulate with apiculate apexes. The lip has no lateral lobes, but there is lip lateral to the callus, just not a distinct separate lobe from the apical lobe. The margins of the apical lobe recurve severely. The callus has about 20 keels that run longitudinally, not in a fan pattern. The callus has a depressed area in its mid portion, rising apically again, the keels become less distinct and form a sulcus before the main apical lobe of the lip. There are no teeth or points on the apical portion of the keels. The column is broad, covering the callus laterally but not apically. The clinandrium is on the ventral side of the column but is still visible from the front. The column is glabrous and has a wide base before the column foot. PLATE 107.

MEASUREMENTS Leaves 25–30 cm long, 4 cm wide; inflorescence 10 cm long; flower 10 cm long, 9 cm wide; dorsal sepal 5.2 cm long, 3.1 cm wide; lateral sepals 5.7 cm long, 3.3 cm wide; petals 4.6 cm long, 3.2 cm wide; lip 4 cm long, 2.2 cm wide.

ETYMOLOGY Greek *koilos*, cavity, probably referencing the sulcus after the callus or the ventral side of the column.

DISTRIBUTION AND HABITAT Known from Colombia, on the Ecuadorian border, at 1000 m in elevation; common in cultivation.

FLOWERING TIME Not recorded.

Pescatorea coronaria Reichenbach f., *Linnaea* 41: 108. 1887. Type: Colombia (holotype: W).

DESCRIPTION The inflorescence is laxly pendent. The flowers are purple-red with white apexes, the lip is a lighter purple or pink, the callus is keeled dark purple-red, and the column is dark purple-red. The sepals are elongated or spatulate with blunt apexes. The three-lobed lip has coarse warty asperities on the front of the callus and numerous long hairs on the convex apical lip. The callus keels remain distinct not coalescing centrally, forming a crescent shape with 21 keels. PLATE 108.

MEASUREMENTS Leaves 35 cm long, 2.7–4.3 cm wide; inflorescence 7 cm long; sepals and petals 3.7 cm long; lip 3 cm long, 2.5 cm wide; column 2.2 cm long.

ETYMOLOGY Latin *cornonaris*, suitable for garlands, referring to the royal color of the flower.

DISTRIBUTION AND HABITAT Known from Colombia, at 1250–1900 m in elevation.

FLOWERING TIME Not recorded.

Pescatorea dayana Reichenbach f., *Gard. Chron.*: 1618. 1872. Type: Colombia (holotype: W).

SYNONYMS
Zygopetalum dayanum Reichenbach f., *Gard. Chron.*: 1618, 1872, *in syn*.
Pescatorea dayana var. *splendens* Reichenbach f. *Gard. Chron.*: 575. 1873.
Pescatorea dayana var. *rhodacra* Reichenbach f., *Gard. Chron.* 2: 226. 1874.
Pescatorea dayana var. *roezlii* Stein, *Orchideenbuch*: 495. 1892.
Zygopetalum dayanum var. *rhodacrum* (Reichenbach f.) F. T. Hubbard.
Pescatorea dayana var. *candidula* Reichenbach f., *Gard. Chron.*: 226. 1874.
Pescatorea dayana subsp. *candidula* (Reichenbach f.) Fowlie, *Orch. Digest* 32: 86–91. 1968.
Pescatorea dayana subsp. *splendens* (Reichenbach f.) Fowlie, *Orch. Digest* 32: 86–91. 1968.

DESCRIPTION The inflorescence is laxly pendent. The sepals and petals are white tipped with green, the petals and lip are white, the callus is purple, becoming more violet apically, and the column is yellow with a red patch at the base. The sepals are broadly ovate, the petals rounded. The rhomboid or quadrate lip is lightly convex. The callus is crescent-shaped, with raised central keels that cause the callus to look like a lowercase *w*. The narrow column has basal auricles laterally, but they are small. PLATE 109.

COMMENT The species is variable in color with many described varieties:

   var. *candidula*, Colombia, sepals and petals pure white, with tint of crimson on lip.
   var. *roezlii*, Colombia, sepals and petals white with violet or rose-purple flecks, lip yellow-violet or purple-red.
   var. *rhodacra*, Panama, white sepals and petals tipped with rose, lip white suffused with crimson, column white with a crimson tip.
   var. *splendens*, dark violet blotches at tips of sepals and petals, lip and base of column deep violet.

Pescatorea dayana differs from *Pescatorea wallisii* in that the former has a less prominent raised central keel, the whole callus forms a shallow lowercase *w*, and the lip skirt is more square, as opposed to the prominent central keel and point of *P. wallisii*, the distinct uppercase *W* the whole callus forms from the front, and the lip skirt that is long and thin and markedly curled at the lateral edges.

MEASUREMENTS Leaves 22–25 cm long; flower 8 cm long, 6.8 cm wide; dorsal sepal 3.5 cm long, 2.4 cm wide; lateral sepals 3.6 cm long, 2.6 cm wide; lip 2.8 cm long, 1.7 cm wide; column 1 cm long.

ETYMOLOGY Named for John Day.

DISTRIBUTION AND HABITAT Known from Colombia, Costa Rica, and Panama.

FLOWERING TIME Late autumn.

Pescatorea ecuadorana (Dodson) Dressler, *Lankesteriana* 5(2): 95. 2005. Basionym: *Bollea ecuadorana* Dodson, *Selbyana* 7(2–4): 354. 1984. Type: Ecuador, El Oro, collected by P. Morgan, 27 July 1979, *M. E. Fallen & C. H. Dodson 8568* (holotype: SEL).

DESCRIPTION The inflorescence is laxly pendent. The sepals and petals are white, densely, almost completely, suffused with pink or purple, more densely on the apical half, the sepal and petal tips are white, the lip is white or light purple, the callus is yellow, and the column is pink or purple. The lip is truncate and its margins curl under. The 12-keeled callus has an upper surface that is basally concave then convex, and rolls into a sulcus of the lower lip, the outer surface of the callus is smooth. The column undersurface is flat with slight wings at the margins. PLATE 110.

COMMENT *Pescatorea ecuadorana* is very similar to *P. coelestis*, but it is generally a less impressive flower with fewer keels in the callus, a shorter lip, and a lighter blue color. Both species are very variable in shape and color; some specimens have very full sepal and petals,

others have thin segments. Since *P. coelestis* is found near the border of Ecuador and Colombia, it most probably is very closely related to *P. ecuadorana*.

MEASUREMENTS Leaves 25–30 cm long, 3.5 cm wide; inflorescence 8 cm long; dorsal sepal 3.5 cm long, 2 cm wide; lateral sepals 4 cm long, 2.2 cm wide; petals 3.4 cm long, 2.2 cm wide; lip 2.5 cm long, 2 cm wide; column 2.2 cm long, 1.4 cm wide.

ETYMOLOGY Named for the country where the species is found.

DISTRIBUTION AND HABITAT Known from Ecuador.

FLOWERING TIME July.

Pescatorea hemixantha (Reichenbach f.) Dressler, *Lankesteriana* 5(2): 95. 2005. Basionym: *Bollea hemixantha* Reichenbach f., *Gard. Chron.* ser. 3, 4: 206. 1888. Type: Colombia (holotype: W).

SYNONYM
Bollea cardonae Schnee, *Rev. Fac. Agric.* 1: 203. 1953. Type: Venezuela.

DESCRIPTION The inflorescence is laxly pendent to upright. The sepals and petals are white, sometimes the petals have a pink-and-yellow flush, the lip is light yellow, the callus is dark yellow, and the column is white, sometimes tinted pink. The dorsal sepal is upright, the petals and sepals are elliptical, much longer than wide. The lip is trullate, the lateral lip margins fold downward, some specimens have only a narrow rim of lateral lip when seen from above, and the apical lip lobe is curled under the callus. The base of the lip before the callus has dense warty asperities and hairs. The callus has 16 keels and consists of half the lip length. There is a flat portion (shelf) in the center of the callus, laterally the keels are palpable, and the keels have a toothlike structure at their apical ends. The large bulky column, appearing like a parrot beak, covers the callus laterally and from above. PLATE 111.

MEASUREMENTS Leaves 30 cm long; inflorescence 5 cm long; sepals 3.5 cm long, 2 cm wide; petals 3 cm long, 1.5 cm wide; lip 3 cm long, 2 cm wide; column 2.5 cm long.

ETYMOLOGY Latin *hemi*, half, and Greek *xantho*, yellow, referring to the color of the petals.

DISTRIBUTION AND HABITAT Known from Colombia and Venezuela, at 100–1300 m in elevation.

FLOWERING TIME Not recorded.

Pescatorea hirtzii (Waldvogel) Dressler,
Lankesteriana 5(2): 95. 2005. Basionym: *Bollea hirtzii* E. Waldvogel, *Orchidee (Hamburg)* 33(4): 143. 1982. Type: Ecuador, Pastaza, Río Topo, 1500 m, leg. A. Hirtz, *Herb. E. Waldvogel 14/81* (holotype: KIEL).

DESCRIPTION The inflorescence is laxly pendent to upright. The sepals, petals, and lip are pale pink-yellow almost white, the column is dark pink to red. The sepals and petals are ovate to elliptic, sometimes with uniform wavy edges. The recurved lip is acuminate at the apex. The callus is ovate, higher in front (apically), with 15 to 17 narrow keels extending to an elevated apex, and the keels of the callus have only a minimal depression between them. The column is thick and broadly winged, hirsute from the middle to the base. The column base has lateral auricles. PLATE 112.

MEASUREMENTS Leaves 60 cm long, 5 cm wide; inflorescence 15–20 cm long; dorsal sepal 3.5 cm long, 2.1 cm wide; lateral sepals 4 cm long, 2.2 wide; petals 3.1 cm long, 1.7 cm wide; lip 3 cm long, 1.5 cm wide; column 2.0–2.4 cm long, 2 cm wide.

ETYMOLOGY Named for Alexander Hirtz.

DISTRIBUTION AND HABITAT Known from eastern Colombia, Ecuador, and Peru, at 1300–1500 m in elevation, in lower montane cloud forest.

FLOWERING TIME February to May.

Pescatorea klabochorum Reichenbach f.,
Gard. Chron. 1: 684. 1879. Type: Colombia (holotype: W).

SYNONYMS
Zygopetalum klabochorum Reichenbach f., *Gard. Chron.* 1: 684. 1879, *in syn.*
Pescatorea fimbrata Regel, *Act. Hort. Petrop.* 6: 293. 1879; and *Gartenfl.* 129, plate 1008. 1880.
Zygopetalum fimbriatum (Regel) N. E. Brown, *Suppl. Johnson's Gard. Dict.*: 1026. 1882.
Pescatorea vervaeti hort. ex T. Moore, *Fl. & Pom.*: 10. 1883.

DESCRIPTION The inflorescence is laxly pendent. The sepals and petals are white, the tips are chocolate-purple or maroon, the lip is yellow (sometimes white) with purple-tipped short hairs and warts, the callus is sulfur-colored with brown (red-maroon) keels, and the column is a dull yellow tinged brown and purple. The sepals are blunt at the apex; the petals have more of a point. The lip is three-lobed, trowel-shaped, the lateral lobes (auricles) are erect, touching the column. The apical lobe of the lip is covered with short hairs and/or warts, which are very stiff, giving the feel of sand paper. The apical lip lobe rolls under at its lateral edges. The callus has approximately 20 distinct keels each with an apical tooth. The column is hirsute on the ventral surface, with a square base, without lateral auricles. PLATE 113.

COMMENT The hairs on *Pescatorea klabochorum* are as stiff as the warts on *P. cerina* in feel; however, they are longer, which is why they are described as hairs. Some authors list *P. fimbriata* as a synonym of *P. lehmannii*, but based on the type drawing, I feel it belongs to *P. klabochorum*.

Two forms are described:

var. *burfordiensis*, lip apex tipped with triangular maroon areas
var. *ornatissimum*, petal tips mauve-purple with mauve-purple spots at the base of the petals, dorsal sepal with a spot at base.

MEASUREMENTS Leaves 30 cm long; flowers 7–10 cm wide; dorsal sepal 4.2 cm long, 2.5 cm wide; lateral sepals 4 cm long, 2.7 cm wide; petals 3.5 cm long, 2.2 cm wide; lip 2.7 cm long, 2.5 cm wide; column 2.8 cm long.

ETYMOLOGY Named for the Klaboch brothers of Germany, who collected in Colombia for Sander's nursery; *klabochora* is the singular form, *klabochorum* the plural form of the name, honoring a family or group.

DISTRIBUTION AND HABITAT Known from Colombia and Ecuador, at 1300 m in elevation.

FLOWERING TIME July.

Pescatorea lalindei (Reichenbach f.) Dressler, *Lankesteriana* 5(2): 95. 2005. Basionym: *Bollea lalindei* Reichenbach f., *Gard. Chron.* 2: 33. 1874. Type: Colombia (holotype: W).

SYNONYMS
Batemannia lalindei Linden, *Cat. Gén.*: 90. 1873.
Zygopetalum lalindei Reichenbach f., *Gard. Chron.* 2: 33. 1874, *in nota*.
Bollea patini Reichenbach f., *Gard. Chron.* 2: 34. 1874.
Zygopetalum patini Reichenbach f., *Gard. Chron.* 2: 84. 1874, *in nota*.

DESCRIPTION The inflorescence is upright. The flower color is dark pink, the column is darker pink, purple, or lavender, and the callus is pink with a hint of yellow. The sepal and petals are reflexed back apically. The petals are narrower than the sepals. The lip reflexes back apically to the callus sulcus, but not much more than 90 degrees, the lip apex is narrow (spade-shaped), and the lip does not extend lateral to the callus. The callus has 14–16 keels, each one blunted at the apical end forming blunt teeth, the teeth of the medial keels form a fused ridge apically. PLATE 114.

COMMENT The petals much narrower than the sepals, the lack of lateral lobes on the lip, and the fused ridge of teeth on the apical end of the callus are diagnostic for *Pescatorea lalindei* and distinguish it from *P. violacea*, which has sepals and petals almost the same width, a portion of the lip that is lateral to the callus, and a callus with the central three or four keels forming teeth which remain distinct and unfused apically.

MEASUREMENTS None recorded.

ETYMOLOGY Named for Don Juan Lalinde, who reportedly was the first Antioqueñan to become interested in orchid collection and cultivation.

DISTRIBUTION AND HABITAT Known only to Colombia, in the Department of Santander.

FLOWERING TIME Not recorded.

Pescatorea lamellosa Reichenbach f., *Gard. Chron.* 2: 225. 1875. Type: Colombia, *Wallis 387* (holotype: W).

SYNONYM
Zygopetalum lamellosum Reichenbach f., *Gard. Chron.* 2: 225. 1875, *in nota*.

DESCRIPTION The inflorescence is laxly pendent to semiupright. The flower is cream-white to pink, the lip is white, the callus is yellow with dark red-brown veins, the column is white-yellow with maroon-red longitudinal veining, and the anther cap is maroon-red. The one-lobed lip margin is revolute with a stiff warty surface on the apical half. The callus of approximately 20 keels extends over half the length of the lip. PLATE 115.

COMMENT *Pescatorea lamellosa* differs from *P. cerina* in that the tepal color is typically more yellow in the former, the lip is whiter, the lamellae are a darker brown, and the lip lacks the lateral auricle-like lobes of *P. cerina*.

MEASUREMENTS Leaves 30–40 cm long; inflorescence 10 cm long; dorsal sepal 3.5 cm long, 2.3 cm wide; lateral sepals 3.5 cm long, 2.5 cm wide; petals 3.2 cm long, 1.7 cm wide; lip 3 cm long, 2.5 cm wide; column 3 cm long, 0.7 cm wide.

ETYMOLOGY Latin *lamellatus*, composed of layers or plates, describing the lip and callus.

DISTRIBUTION AND HABITAT Known from Colombia and Ecuador, at 1000–2000 m in elevation, in areas of constant high humidity.

FLOWERING TIME Not recorded.

Pescatorea lawrenceana (Reichenbach f.) Dressler, *Lankesteriana* 5(2): 95. 2005. Basionym: *Bollea lawrenceana* Reichenbach f., *Gard. Chron.* 2: 266. 1878. Type: Colombia (holotype: W).

SYNONYM
Zygopetalum lawrenceanum Reichenbach f., *Gard. Chron.* 2: 266. 1878, *in nota*.

DESCRIPTION The inflorescence is laxly pendent. The sepals and petals are white to pink to reddish purple with a band of dark magenta (purple-red) at the mid point to the apex, the lip has a magenta apical lobe and yellow callus, and the column is purple. The lip as it emerges off the column foot is held more vertical and higher than most of the group of *Bollea*, so that the column foot is often visible below the lip as viewed from the front. The lip has an elongate recurved acuminate apex. The callus has 12–14 keels. The column is narrow at the base with basal lateral auricles, broadly expanding at the apex to cover the callus. Stunning color forms have yellow on the lower half of the lateral sepals. PLATE 116.

COMMENT *Pescatorea lawrenceana* is similar to *P. pulvinaris* but can be distinguished by its callus, which is more oblong (longer than wide) compared to the huge square-shaped (as wide as long) callus of *P. pulvinaris*. *Pescatorea lawrenceana* also has basal auricles on its column, *P. pulvinaris* does not. See *P. pulvinaris* for further details.

MEASUREMENTS Leaves 40 cm long, 5 cm wide; inflorescence 25 cm long; dorsal sepal 4 cm long, 1.7 cm wide; lateral sepals 4 cm long, 2 cm wide; petals 4 cm long, 1.5 cm wide; lip 3 cm long, 2 cm wide; column 2.5 cm long, 2 cm wide.

ETYMOLOGY Named for Mr. Lawrence.

DISTRIBUTION AND HABITAT Known from Colombia and Ecuador, at 1200–1500 m in elevation, in extremely wet lower montane forest.

FLOWERING TIME Sporadically most of the year.

Pescatorea lehmannii Reichenbach f., *Gard. Chron.* 2: 424. 1879. Type: Colombia (holotype: W).

SYNONYMS
Zygopetalum lehmannii Reichenbach f., *Gard. Chron.* 2: 424. 1879, *in syn.*
Pescatorea lehmannii var. *magnifica* Cogniaux, *Chron. Orch.*: 377. 1903.

DESCRIPTION The inflorescence is laxly pendent. The sepals and petals are white with red purple apexes. The species is noted for the white veins that extend to near the apexes, giving the appearance of close, longitudinal, parallel, wide lines of red-purple (sometimes pink with red lines). The lip is red-purple (pink spotted with red) with white hairs, the callus is chestnut brown, maroon, or purple. The sepals and petals are ovate. The lip is weakly three-lobed, with small lateral lobes that recurve back, more just a fringe of tissue off the lateral sides of the callus. The lip narrows at its apex, and the lip is clothed in long stiff hairs on its upper surface, arranged on the elevated veins of the lip. The callus is an inverted *M*, with a midline elevated keel formed by the fusion of the central keels that begins at the base of the lip, the lateral keels are more shallow until they reach the fold in the callus, where they become more raised. The flower is weakly fragrant of citrus. PLATE 117.

COMMENT The color of this species is variable, sometimes because of hybrids labeled as the species.

MEASUREMENTS Leaves 30–40 cm long, 3 cm wide; inflorescence 12 cm long; dorsal sepal 4 cm long, 3 cm wide; lateral sepals 4.4 cm long, 3.8 cm wide; petals 3.8 cm long, 2.8 cm wide;

lip 2.4–4.0 cm long, 2 cm wide; column 2 cm long.

ETYMOLOGY Named for F. C. Lehmann, a collector in Colombia and the German consul in New Granada.

DISTRIBUTION AND HABITAT Known only from Colombia and Ecuador, at 1000–1400 m in elevation, in extremely wet montane cloud forest.

FLOWERING TIME Most of the year.

Pescatorea pulvinaris (Reichenbach f.) Dressler, *Lankesteriana* 5(2): 95. 2005. Basionym: *Bollea pulvinaris* Reichenbach f., *Linnaea* 41: 107. 1877. Type: Colombia, *Wallis s.n.* (holotype: W).

SYNONYM
Zygopetalum pulvinare Reichenbach f., *Linnaea* 41: 107. 1877, *in syn.*

DESCRIPTION The inflorescence is laxly pendent. The base color of the flower is red-purple, the sepals and petals often have white apexes, sometimes light purple medially becoming darker midway, almost seeming to be like bands of red-purple, the lip apex is a velvet red-purple, the callus is yellow overlaid with red-purple, and the column is red-purple, with the ventral surface being yellow basally with red spots. The lip is markedly concave, to the point that sometimes all that is visible from the front is the callus. The lip apex is pointed, not notched. The callus is broad, extending to the lateral edges of the lip, and composed of 12–16 warty keels. The column barely covers the entire callus width, lacks lateral auricles on its base, and is hirsute on its ventral surface. PLATE 118.

COMMENT This is probably the biggest bulkiest flower of this genus; other flowers may be wider, but their substance doesn't match that of *Pescatorea pulvinaris*. Colombian growers Francisco Villegas and Gustavo Aguirre (pers. comm.) have pointed out that there seemed to them to be two forms of *P. pulvinaris*. One form opens fully, spreading the sepals and petals till they are perpendicular to the column and lip. The other form remains cupped when mature. When I was in Colombia, plants of both forms were in bloom at the same time and they are different species. The cupped-shaped flower has auricles at the base of the column and its callus is a little longer than wide, which would make the cupped specimens *P. lawrenceana* with the coloration of a "typical" *P. pulvinaris*.

Pescatorea pulvinaris is very similar to *P. lawrenceana* but can be distinguished by its callus, which is huge and square-shaped (as wide as long) compared to the more oblong (longer than wide) lip callus of *P. lawrenceana*. *Pescatorea pulvinaris* also lacks the basal lateral auricles on the column, which are distinctive on *P. lawrenceana*.

MEASUREMENTS Leaves 60 cm long, 3–4 cm wide; inflorescence 10–15 cm long; sepals 4 cm long, 3 cm wide; petals 3.8 cm long, 2.5 cm wide; lip 4 cm long, 2.1 cm wide (opened up); column 2.7 cm long, 1.8 cm wide.

ETYMOLOGY Latin *pulvinatus*, cushion-shaped, strongly convex, referring to the shape of the lip.

DISTRIBUTION AND HABITAT Known only from Colombia, at 1800–2500 m in elevation.

FLOWERING TIME Not recorded.

Pescatorea violacea (Lindley) Dressler, *Lankesteriana* 5(2): 95. 2005. Basionym: *Huntleya violacea* Lindley, *Sert. Orchid.*: plate 26. 1839; *Bot. Reg.* 25: misc 17. 1839. Type: Guyana, Demerara (holotype: K), type for the genus *Bollea*.

SYNONYMS
Bollea guianensis Klotzsch, *Schomb. Fauna & Fl. Guy.*: 1206. 1841. *nomen.*
Bollea violacea (Lindley) Reichenbach f., *Bot. Zeit. (Berlin)* 10: 668. 1852.
Zygopetalum violaceum (Lindley) Reichenbach f., *Ann. Bot. Syst.* 6: 650. 1863.

DESCRIPTION The inflorescence is laxly pendent. The flowers are purple to violet, rarely the sepals and petals are partially yellow, the lip and column are light brown basally, deep violet apically, and the callus is yellow-violet. The lip is concave basally developing into a transverse ridge, becoming sharply convex in the area of the callus, and then folding after the callus in a concave shape to form a pleat or sulcus. The apical lip margins inroll. A portion of the lip extends lateral to the callus. The callus has 11–13 keels that are toothed apically, the medial three form very prominent teeth or knobs and not are fused. The column covers the callus from above, resembling a parrot's beak. PLATE 119.

MEASUREMENTS Leaves 12–70 cm long, 2–5 cm wide; inflorescence 6–16 cm long; dorsal sepal 3.0–3.7 cm long, 1.7–2.0 cm wide; lateral sepals 3.0–3.7 cm long, 2.3–3.0 cm wide; petals 3–3.7 cm long, 1.4–1.8 cm wide; lip 3.0–3.2 cm long, 2.2–2.6 cm wide; column 2.5–3.0 cm long, 2 cm wide.

ETYMOLOGY Latin *violaceus*, red-blue colors nearer the blue, referring to the flower.

DISTRIBUTION AND HABITAT Known from northern South America including northern Brazil, recorded at 450 m in elevation.

FLOWERING TIME Not recorded.

Pescatorea wallisii Linden & Reichenbach f., *Gard. Chron.*: 710. 1869. Type: Ecuador (holotype: W).

SYNONYMS
Zygopetalum wallisii Reichenbach f., *Gard. Chron.*: 710. 1869, *in syn.*; Van Houtte, *Fl. des Serres* 18: 17. 1869.
Pescatorea roezlii Reichenbach f., *Gard. Chron.* 1: 755. 1874.
Zygopetalum roezlii (Reichenbach f) Reichenbach f., *Gard. Chron.* 1: 620. 1877.
Zygopetalum euglossum (Reichenbach f.) N. E. Brown, *Suppl. Johnson's Gard. Dict.*: 1026. 1882.
Pescatorea euglossa Reichenbach f., *Gard. Chron.* 2: 808. 1876.

Bollea schroederiana Sander, *Gard. Chron.* ser. 3, 17: 497, fig. 70. 1895, as *B. schroderiana*.
Pescatorea schroederiana (Sander) Rolfe, *Orch. Rev.* 8: 68. 1900. This name is listed as an accepted species on the Kew World Checklist of Monocotyledons.

DESCRIPTION The inflorescence is laxly pendent. The sepals and petals are milk-white gradually developing a light violet-rose suffusion apically, the lip is violet-rose or purple, the callus is ivory white, striped with very soft rose and frequently entirely tinted, and the column is white with rose spots at the base. The petals are sometimes notched at the apex, but otherwise the apex is blunt. The lip is heart-shaped at the base, lacks lateral lobes, and is bilobed at the tip (notched at apex) and glabrous. The callus has 12–17 keels which continue onto the lip as veins. The medial keel is higher than those lateral to it, forming a point in the center of the callus. Laterally the callus again rises, becoming the highest portion of the callus, so the whole callus forms a *W* from the front. The column is hirsute at the base and has basal auricles laterally. The flower has a citrus fragrance. PLATE 120.

COMMENT This is another species that is very variable in coloration from pinks to purple, but the W-shaped callus and the long thin notched lip apex are diagnostic.

MEASUREMENTS Leaves 60 cm long; inflorescence 12–17 cm long; flower 7–10 cm wide; sepals and petals 3 cm long, 2 cm wide; lip 3.5 cm long, 2 cm wide; column 2 cm long.

ETYMOLOGY Named for Gustav Wallis, a collector of Colombian orchids.

DISTRIBUTION AND HABITAT Known from Colombia and Ecuador, at 70–2500 m in elevation.

FLOWERING TIME July to September, to most of the year.

• •

Several natural hybrids or aberrant forms of *Pescatorea* bear mentioning as they are often listed in the literature as species. With further study, a few may turn out to be true species.

Key to *Pescatorea* Including Aberrant Forms and Natural Hybrids

1a. Column with wings that cover the lateral sides of the lip callus, these are (some of) the "old" species of *Bollea* . . . . go to 2

1b. Column does not cover the lateral sides of the lip callus, these are the "old" species of *Pescatorea* . . . . go to 10

2a. Clinandrium completely hidden beneath apex of column, so that the anther cap is not seen from the frontal view, column covers callus from above . . . . go to 3

2b. Clinandrium not completely hidden beneath apex of column, so that the anther cap is visible from the frontal view, callus visible from above . . . . go to 6

3a. Base of lip before callus with dense warty asperities and hairs; sepals and petals white or yellow, inferior half of lateral sepals same color as superior half; callus of 16 keels . . . . *P. hemixantha*

3b. Base of lip before callus without asperities or hairs; sepals and petals purple or pink, callus of 16 or fewer keels. . . . . go to 4

4a. Callus central keels form three distinct teeth at their apex; flower dark purple including lip, callus white, callus with 10 keels . . . .*P. violacea*

4b. Callus central lamellae form a fused ridge at their apex, callus with 14 to 16 keels . . . . go to 5

5a. Flower pink to light purple . . . . *P. lalindei*

5b. Sepals and petals white; lower half or lateral sepals yellow . . . . *P. whitei*

6a. Auricles present on base of column . . . . go to 7

6b. Auricles lacking on base of column . . . . go to 8

7a. Column and callus widest of the genus, 1.8 cm or more wide; lip velvet red apical to salmon-pink-colored callus . . . .*P. pulvinaris*

7b. Column and callus 1–1.5 cm wide, sepals and petals white or pale pink, column dark pink, red, or purple. . . . . *P. hirtzii*

8a. Callus longer than wide, lip with a long point, sepals and petals with purple, column not solid pink or purple. . . . .*P. lawrenceana*

8b. Callus as wide as or wider than long . . . . go to 9

9a. Base of column with acute posterior appendages and acute ventral tooth at base of column, callus with 12 keels . . . .*P. ecuadorana*

9b. Base of column with blunt posterior appendages and blunt ventral tooth at base of column, callus with 20 keels . . . .*P. coelestis*

10a. Leaves 25–30 times longer than wide, long for genus; sepal and petal apexes and lip steel blue, callus gold-yellow. . . . .*P. triumphans*

10b. Leaves 10–12 times longer than wide, typical for genus . . . . go to 11

11a. Labellum with hairs, warts, or bristles . . . . go to 12

11b. Labellum without hairs, warts, or bristles . . . . go to 21

12a. Lip concave, flower red . . . . *P. cochlearis*

12b. Lip convex with lateral margins turned under, flower not red . . . . . go to 13

13a. Lip with hard warts not hairs . . . . go to 14

13b. Lip with warts and hairs . . . . go to 18

14a. Flower color yellow or green . . . . go to 15

14b. Flower with red, lavender, or purple on sepals and petals . . . . go to 17

15a. Callus with 7 keels. . . . . *P. kalbreyeriana*

15b. Callus with more than 10 keels. . . . . go to 16

16a. Lip three-lobed, lateral lobes small, lip yellow with brown lamellae,
     13–15 keels . . . . *P. cerina*

16b. Lip one-lobed, lip apex white with brown lamellae, 13–15 keels *.P. lamellosa*

17a. Sepals and petals white with purple-violet tips, lip cream-colored, callus
     yellow, column purple at base. . . . .
     . . . . *P. backhousiana* or *P. klabochorum* (see descriptions following)

17b. Sepals and petals dark lavender with darker tips, lip cream-colored, callus
     yellow with brown ridges, 15–17 keels. . . . . *P. ×gairiana*

18a. Callus with three central keels almost confluent into one structure . . . . . .
     . . . . go to 19

18b. Callus otherwise than above, sepals and petals with red, lavender, or
     purple somewhere . . . . go to 20

19a. Lateral keels with angles in a row, sepals and petals white with yellow tips,
     no red or lavender, callus yellow. . . . . *P. dormaniana*

19b. Lateral keels without angles, sepals and petals with longitudinal
     striping of red or lavender (could be thought of as white veins on purple
     background) . . . . *P. lehmannii*

20a. Wart and hairs short, lip surface easily seen, sepals and petals white,
     tipped red-lavender, lip cream, callus yellow, lamellae red. . . *P. klabochorum*

20b. Warts and hairs long, densely obscuring lip surface, sepals and petals red
     or purple-red, lip lighter red, callus dark red. . . . . *P. coronaria*

21a. Sepals and petals undulate and pointed, sepals and petals white, tipped
     light lilac, lip purple with some yellow on side lobes, callus white . . . .
     . . . . *P. rueckeriana*

21b. Sepals and petals not markedly undulate or pointed . . . . go to 22

22a. Callus crescent-shaped, with central keel or lamella raised . . . . . . go to 23

22b. Callus not crescent-shaped, without central raised portion. . . . . . . go to 24

23a. Lip roughly square or oval, as long as wide . . . . *P. dayana*

23b. Lip longer than wide, apex heart-shaped . . . . *P. wallisii*

24a. Sepals and petals white with purple flecks, lip brown-purple, apex
     lavender, callus yellow with red lines. . . . . *P. russeliana*

24b. Sepals and petals white-violet with a band of dark purple-violet at the
     apex, lip white- yellow, apex with purple-violet blotch, lip covered by
     numerous purple spots, keels of callus purple . . . . *P. ×bella*

Pescatorea backhousiana Reichenbach f.,

Gard. Chron., n.s., 8: 456. 1877. Type: Ecuador (holotype: W).

SYNONYMS

Zygopetalum backhousianum Reichenbach f., *Gard. Chron.* 2: 456. 1877, *in syn.*

Pescatorea klabochorum var. *backhousiana* (Reichenbach f.) Stein, *Orchideenbuch*: 496. 1892.

DESCRIPTION The sepals and petals are cream-white with tips marked purple-violet, the lip is cream-colored, the callus is yellow with brown interior veins, and the column is purple at the base, yellow-white at the apex. The lateral sepals taper gradually, instead of abruptly at the apex as other species do. The lip is three-lobed, with warty asperities on the apical lobe but no hairs. The callus has 19 keels, but no angular callus at the base of the lip.

COMMENT The flowers are 9 cm across and are also described as being the size of *Pescatorea cerina* flowers. *Pescatorea backhousiana* was described but has not been rediscovered, or if it has been rediscovered, it was identified as belonging to another species. The original pictures look very similar to *P. klabochorum*, and indeed may just be a variety or form of that species. I have left it as a species in this text, not being able to match the descriptions of the two species completely, since *P. klabochorum* has hairs and warts and *P. backhousiana* is hairless.

MEASUREMENTS None.

ETYMOLOGY Named for Mr. Backhouse.

DISTRIBUTION AND HABITAT Known only from Ecuador.

FLOWERING TIME Recorded in June to August.

Pescatorea ×bella (Reichenbach f.) P. A.

Harding, *comb. nov.* Basionym: *Pescatorea bella* Reichenbach f., *Gard. Chron.* 9: 492. 1878. Type: Colombia (holotype: W).

SYNONYM

Zygopetalum bellum Reichenbach f., *Gard. Chron.* 1: 492. 1878, *in nota.*

DESCRIPTION The sepals and petals are light white-violet, apically with a band of dark purple-violet. The lip is white with yellow at the base, a purple-violet blotch, and covered by numerous purple spots. The callus keels are purple. The sepals and petals are narrow and elongate (drawn out), unlike those of *Pescatorea dayana*, which get wider at the apex. The lip is elliptic. The callus is large with 21 keels.

COMMENT This was thought to be natural hybrid by Heinrich Gustav Reichenbach. The picture of the type looks very similar to the picture of *Bollea coelestis* 'Electra' FCC/AOS on the back cover of the *Awards Quarterly* 36(4) and in *Orchids* 75(6): 434, which is misidentified.

MEASUREMENTS Leaves 25 cm long; flowers 10 cm high, 12 cm wide.

ETYMOLOGY Spanish *bello*, beautiful.

DISTRIBUTION AND HABITAT Known from Colombia.

FLOWERING TIME March to May.

Pescatorea dormaniana (Nichols)

Reichenbach f., *Gard. Chron.* 1: 330. 1881. Type: Colombia (holotype: W).

DESCRIPTION The flower is white with light sulfur color on the callus and at end of the sepals. The lip has acuminate papillae (gradually tapering to a point), like *Pescatorea klabochorum* and *P. lehmannii* according to Heinrich Gustav Reichenbach. The callus has a continuous row of angled ridges on the posterior margins of the side laciniae and three prolonged connate median keels. The base of the column is sagittate.

COMMENT This species to my knowledge has not been collected again, and I suspect that it is just an alba form of *Pescatorea lehmannii*, as many of the specimens I have seen with normal *P. lehmannii* coloring have these pointed ridges on the lateral keels of the callus. The drawing of the type of *P. dormaniana* shows a lip very similar to the lip of *P. lehmannii*, having the medial lobe of the lip narrow and covered densely with long hairs. According to Fowlie (1968), the specimen is lost, replaced by a *Warczewiczella*, so the drawing is all that remains.

MEASUREMENTS None recorded.

ETYMOLOGY Named for Charles Dorman, an English orchid enthusiast of the nineteenth century.

DISTRIBUTION AND HABITAT Known only from Colombia, at unrecorded elevations.

FLOWERING TIME Not recorded.

Pescatorea ×*gairiana* (Reichenbach f.)
Fowlie, *Orch. Digest* 32: 89. 1968. *pro sp.*
Basionym: *Pescatorea gairiana* Reichenbach f., *Gard. Chron.* 1: 684. 1879. Type: (holotype: W).

SYNONYMS
Zygopetalum gairianum (Reichenbach f.) Reichenbach f. ex B. D. Jackson, *Index Kew.* 2: 1255. 1895.
Pescatobollea bella Rolfe, *Orch. Rev.* 13: 329. 1905.

DESCRIPTION The sepals and petals are dark violet, the apexes are darker violet, the lip is purple-red or purple-brown, the callus is orange-yellow streaked with purple or crimson, and the column is white with purple spots at the base. The sepals and petals are ribbon-shaped. The lip is ovate with a blunt apex, and has obscure longitudinal keels and numerous warts between the keels. The callus has teeth or warts before the lip flexes backwards, and 15–17 keels.

COMMENT This is a natural hybrid between *Pescatorea coelestis* and *P. klabochorum*. Carlos Uribe Vélez sent me a picture of *P.* ×*gairiana*, which shows ears (lateral lobes) and warts on the lip like those of *P. klabochorum*, and with a color pattern like *P. coelestis* only blue-maroon instead of blue.

MEASUREMENTS Leaves 12–20 cm long; inflorescence 10 cm long; flowers 10 cm wide.

ETYMOLOGY Named for Mr. Gair, who owned the original plant.

DISTRIBUTION AND HABITAT Colombia and Ecuador, at 1300 m in elevation.

FLOWERING TIME Not recorded.

Pescatorea kalbreyeriana Kraenzlin,
Notizbl. Bot. Gart. Berl. 7: 429. 1920, *in syn.*

SYNONYM
Zygopetalum kalbreyerianum Kraenzlin, *Notizbl. Bot. Gart. Berl.* 7: 429. 1920. Syntype: Colombia, Antioquia, near Dos Quebrados, in shady forest, 1400–1500 m, *M. Kalbreyer 1357* (syntype: W); near Frontino, 1700 m, *M. Kalbreyer 1901* (syntype: W).

DESCRIPTION The flowers are green-yellow, the white lip has a purple-violet suffusion with brown marks. The dark green shiny leaves are oblanceolate, rather shortly acute at their apexes. The sepals are elliptic, obtuse, and beautifully reticulate. The petals are obovate-oblong, much smaller than the sepals, narrowing toward their base. The lip has a long linear claw at the base. The lip base is covered with minute tubercles. The blade of the lip has mostly straight lines at the base, otherwise it is entire and broadly oblong, rounded in front, and strongly convex. The entire apical lip margin is involute. The callus is minutely papillose with seven short keels produced evenly in a semicircle. The column is slender and appears smooth.

COMMENT Eric Christenson provided translation of the Latin description of this species, which I have used as the basis for the above description. The German text that accompanies the description reinforces the Latin text. Interestingly, Friedrich Kraenzlin in the text said he didn't really like the flower. When this reference first surfaced, thank you Eric, my initial reaction was that there was no *Pescatorea* with seven keels. I received a picture of a flower that was in the collection of Francisco Villegas of Medellín, Colombia. Francisco doesn't know where the plant came from other than Colombia, and always thought it was an aberrant *P. lamellosa*. Now it is dead. Was it *P. kalbreyerianum*? It was an excellent match. Without seeing the type specimen, it is hard to reach a conclusion. Just another mystery left to be solved.

MEASUREMENTS Leaves 50 cm long, 4 cm wide; inflorescence 8 cm long; sepals 3.5 cm long, 1.7–1.8 cm wide; petals 3 cm long, 1 cm wide; lip 2 cm long, 1.2 cm wide; column 1.8 cm long.

ETYMOLOGY Named for M. Kalbreyer.

DISTRIBUTION AND HABITAT If this species really exists, it is from Colombia.

FLOWERING TIME Not recorded.

Pescatorea ×pallens (Reichenbach f.) P. A. Harding, *comb. nov.* Basionym: *Bollea pallens* Reichenbach f., *Gard. Chron.* 1: 462. 1881.

SYNONYM
Pescatobollea ×pallens (Reichenbach f.) Fowlie, *Orch. Digest* 33: 103. 1969.

DESCRIPTION The column is much reduced in size, smaller than a normal *Bollea*, barely larger than the old definition of *Pescatorea*, and does not fully conceal the lip from above.

COMMENT According to Fowlie (1969a), this is a natural hybrid.

Pescatorea rueckeriana Reichenbach f., *Gard. Chron.* 2: 424. 1885. Type: Colombia (drawing: W).

DESCRIPTION The sepals and petals are cream-white tipped with light lilac, also described as white with green apexes with a large light purple area near the apexes; the lip is purple with some yellow on the side lobes, and the callus is white. The sepals and petals are very acute, undulate, and twisted. There are no bristles on the lip.

COMMENT Fowlie (1968) suggested that this plant belongs to *Pescatobollea* (now a *Pescatorea* hybrid, using the most recent definition of *Pescatorea*), a natural hybrid. Looking at the drawing of the type, it certainly looks like a hybrid, or an aberrant *P. wallisii*, as the callus and column are drawn similar to *P. wallisii*; however, the apical lip is more revolute than that of *P. wallisii* and the tepals are very long and narrow unlike *P. wallisii*. There have been no recent specimens or collections of the plant. The Kew World Checklist of Monocotyledons has this as an unplaced name, meaning that it is undetermined whether it is a true species, a synonym of another species, or a hybrid.

MEASUREMENTS None recorded.

ETYMOLOGY Named for Joseph Ruecker of Venezuela.

DISTRIBUTION AND HABITAT None given.

FLOWERING TIME Not recorded.

Pescatorea russeliana Reichenbach f., *Gard. Chron.* 2: 524. 1878. Type: Ecuador (holotype: W).

SYNONYM
Zygopetalum russelianum Reichenbach f., *Gard. Chron.* 2: 524. 1878, *in syn.*

DESCRIPTION The sepals and petals are cream-white with red-purple flecks, washed or tipped with lavender or red-purple, the lip is brown-purple with lavender at the apex, the callus is yellow with carmine-red lines, and the column is yellow with a brown base. The sepals and petals are ovoid and dilate just before their apexes. The petals are not undulate. The callus has 15 keels with a prominent central keel, the middle keel is straight, the other keels angle or fan out.

COMMENT In the original description, Heinrich Gustav Reichenbach says that this species looks like *Pescatorea roezlii* (which is considered a synonym of *P. wallisii*) except that it has a lip with keels that spread out at angles like the keels of

P. lamellosa. There have been no recent specimens or collections of the plant. The drawing that accompanies the type specimen could certainly fit *P. wallisii*, and the column especially fits that species. The drawing shows keels that don't spread out but appear to all be in the same plane, but it also doesn't depict a prominent central keel or the inverted M-shaped callus which is so typical of *P. wallisii*.

MEASUREMENTS Inflorescence 6–8 cm long; flowers 10 cm wide.

ETYMOLOGY Named for J. Russell of Falkirk, who flowered the type specimen.

DISTRIBUTION AND HABITAT Recorded as being from Ecuador, at unrecorded elevations.

FLOWERING TIME Not recorded.

Pescatorea triumphans Reichenbach f. & Warszewicz, *Bonplandia* 2: 97. 1854. Type: Colombia (holotype: W).

DESCRIPTION The sepals and petals are white with steel blue apexes, the lip is a lighter steel blue, and the callus is golden yellow. The leaves are linear, 25–30 times as long as wide. The lip is ovate, barely retuse at its apex, and glabrous. The callus occupies the proximal third of the lip and is composed of several lamellae.

COMMENT There have been no recent specimens or collections of the plant. The length of the leaves is from a sketch in *Xenia Orchidacea*. The drawing of the type shows a narrow column with prominent auricles at the base of the column and a lip that has no midline prominent keel. The most notable feature of this species is the long thin leaves, but whether this is real or an artist's aberration is difficult to say. This species could be a synonym of *Pescatorea dayana*, especially one of the forms with wide lips, but examination of the type would be necessary.

MEASUREMENTS None given.

ETYMOLOGY Greek *thriambos*, hymn to Dionysus, a public spectacle, or perhaps exalting, referring potentially to the flower or the leaves.

DISTRIBUTION AND HABITAT Known only from Colombia, at unrecorded elevations.

FLOWERING TIME Not recorded.

Pescatorea whitei (Rolfe) Dressler, *Lankesteriana* 5(2): 95. 2005. Basionym: *Zygopetalum whitei* Rolfe, *Gard. Chron.* 3(7): 354. 1890. Type: Colombia (holotype: K).

SYNONYM
Bollea whitei (Rolfe) Schlechter, *Repert. Spec. Nov. Regni Veg. Beih.* 7: 269. 1920.

DESCRIPTION The dorsal sepal and petals are white or cream-white, the lateral sepals are two-toned, the upper half cream-white, the lower half yellow, the lip and callus are yellow, and the column has a white dorsal surface and is yellow basally on the ventral surface becoming white apically. There are warts on the lip basally. The callus has 14–16 keels. The column covers the entire callus.

COMMENT Neudecker (1982) has an article about this plant and a photograph; he discusses this as being a possible alba form of *Pescatorea/Bollea lawrenceana* or *B. lalindei*. In the same publication and issue, Waldvogel (1982) wrote an article on *Bollea* and *Pescatorea*; his picture of *P. lalindei* on page 143 (labeled *Bollea lalindei*) has the same form and characteristics excepting the color as Neudecker's specimen on page 129. As there is no specimen, this mystery may never be solved.

MEASUREMENTS Leaves 25–35 cm long, 3.5–4.5 cm wide; column 1.9 cm wide.

ETYMOLOGY Named for Mr. White.

DISTRIBUTION AND HABITAT Known from Colombia, at unrecorded elevations.

FLOWERING TIME Not recorded.

Stenia Lindley, *Bot. Reg.* 23: plate 1991. 1837.

John Lindley created the genus *Stenia* in 1837, recognizing the particular traits of *S. pallida*—broadly lanceolate leaves and flowers with rigid saccate lips—as being unlike the traits of other genera known at the time. *Stenia guttata* was added to the genus in 1880. The two species remained the only described *Stenia* species for many years until *S. saccata* was described in 1969. *Stenia saccata* was removed from *Stenia* in 1979 by James Ackerman to form the genus *Dodsonia* along with *D. falcata*.

The 1980s and 1990s brought many more species described within the genus *Stenia*. *Stenia glatzii* was added in 2000 and is thought to be a morphological intermediate between *Stenia* and *Dodsonia*. Molecular data neither separate nor combine the two genera of *Stenia* and *Dodsonia*, but since there is now this intermediary, it seems prudent to combine the two genera into one genus, *Stenia*, which is what Dressler et al. did in 2005.

Plants of *Stenia* are epiphytes, growing in upright tufts. The plants lack pseudobulbs or a distinct stem. The leaves get progressively larger with each new leaf, are arranged in a fan pattern, and are green and not papillose. The shape of the leaves is characteristically oblanceolate, conduplicate with some species having a subpetiolate base (having a leaf stalk base) which articulates to the distichous, imbricate green sheaths. The flowers are produced one per inflorescence and are laxly pendent, so that they often rest on the surface of the media if grown in a pot. The inflorescence has small bracts. The flowers are fairly open, the sepals and petals spreading in most species. The petals are decurrent on the column base and the lateral sepals attach to the prominent column foot. The lip is very rigid, fleshy, and deeply concave with a broad laminar or keeled, often ornate callus. The column is straight in some species, curved in others, and the column hoods over the pollinia, the pollinia being on the ventral side of the column. The column wings are on the lateral side of the column though sometimes they are only minimal ridges, and in most species, the column foot is at right angles to the column. The four pollinia are variable. The stipe is very prominent in some species and often fimbriate or toothed on its sides.

Stenia species are found throughout tropical and semitropical South America. More species are found in the Andes, as one would expect, as there are more species of most genera in the Andes.

Stenia is easy to distinguish from *Benzingia*, the other genus in the *Huntleya* alliance that forms lips that could be considered saclike. Most *Benzingia* species have an acute apical point like a tail on the lip apex, whereas *Stenia* species lack this tail and form an apical point by the folding of the apical lip. The leaves of *Stenia* are green and not papillose, those of *Benzingia* are glaucous gray-green. *Stenia*, *Benzingia*, and *Daiotyla* fall in the same group on the parsimonious tree of molecular data, but interesting enough, *Stenia* is closer to *Daiotyla* than *Benzingia*.

Etymology

Greek *stenos*, narrow, suggestive of the characteristic long narrow pollinia of the type species.

List of the Species of *Stenia*

Stenia angustilabia

Stenia aurorae

*Stenia bismarckii**

Stenia bohnkiana

*Stenia calceolaris**

Stenia christensonii

*Stenia glatzii**

Stenia guttata

Stenia jarae

Stenia lillianae

Stenia luerorum

Stenia nataliana

*Stenia pallida** (type species of *Stenia*)

Stenia pastorellii

Stenia pustulosa

*Stenia saccata**

Stenia stenioides

Stenia uribei

Stenia vasquezii

*Stenia wendiae**

* Molecular sampling confirms placement of this species within the genus *Stenia*.

Key to the Species of *Stenia*

1a. Lip with the saccate portion extending more forward than the apex of the midlobe, lateral lobes of lip clasp column (*Dodsonia sensu stricto*). . . . go to 2

1b. Lip with the most forward-extending portion being the apex of the midlobe, the midlobe concave, forming a beak (*Stenia sensu stricto*) . . . . . go to 3

2a.Callus plate ovate with a truncate apex, without any lateral projections . . . . *S. falcata* (see *S. saccata*)

2b. Callus plate with conspicuous pincerlike lateral projections. . . . . *S. saccata*

3a. Margin of lip lobes with definite teeth, tepals beige, lip yellow with fine red dots . . . . *S. glatzii*

3b. Margin of lip lobes smooth without definite teeth. . . . . go to 4

4a. Column with definite bend at the stigma, angle between column and column foot 160 degrees, stipe very long . . . . go to 5

4b. Column without definite bend at the stigma, angle between column and column foot about 90 degrees, stipe variable. . . . . go to 6

5a. Callus teeth blunt, flower green, lip yellow with brown spots on interior surface . . . . *S. jarae*

5b. Callus teeth sharply pointed, flower yellow, lip yellow with brown spots on interior surface . . . . *S. lillianae* ex D. E. Bennett & Christenson

6a. Petals spotted . . . . go to 7

6b. Petals not spotted . . . . go to 10

7a. Lip does not extend as far forward as the column apex (lip extends less forward than column) . . . . *S. calceolaris*

7b. Lip extends more forward than the column apex . . . . go to 8

8a. Lip 2 cm long, 0.7 cm wide, long and thin . . . . *S. guttata*

8b. Lip 1.5 long, more than 1 cm wide, short and stout . . . . go to 9

9a. Lip held so that the apical point is the lowest point of the lip . . . *S. nataliana*

9b. Lip upper margin held horizontally so that lip apex is not the lowest point in the lip . . . . *S. vasquezii*

10a. Lip extends more forward than the column apex . . . . go to 11

10b. Lip extends a length equal to or shorter than the column apex. . . . go to 16

11a. Callus has only three keels with lateral two teeth longest, flower cream yellow . . . . *S. angustilabia*

11b. Callus has more than three keels . . . . go to 12

12a. Callus has five medial keels that don't form a raised area at their apex or extend to lateral walls of lip . . . . *S. uribei*

12b. Callus has more than five keels that extend to lateral walls of lip and form a transverse raised area. . . . . go to 13

13a. Lip lateral edges fold out . . . . *S. pallida*

13b. Lip lateral edges upright or turn in . . . . go to 14

14a. Sepals and petals with long pointed apexes, dorsal sepal much longer than wide. . . . . *S. bohnkiana*

14b. Sepals and petals with blunt apexes, dorsal sepal oval or almost as wide as long . . . . go to 15

15a. Column foot shorter than column, sepals and petals green, lip mahogany-brown. . . . . *S. aurorae*

15b. Column foot longer than column, flower yellow with red-brown on lip . . . . *S. stenioides*

16a. Dorsal sepal oval, nearly as wide as long. . . . . go to 17

16b. Dorsal sepal elliptical, clearly longer than wide . . . . go to 18

17a. Column foot and claw fold into a high arch with a distinctive pointed tip . . . . *S. wendiae*

17b. Column foot and claw fold into a high arch without a distinctive pointed tip . . . . *S. christensonii*

18a. Lip attachment to column not covered by lip lateral folds, column foot nearly same length as column. . . . . *S. bismarckii*

18b. Lip attachment to column foot covered by lip lateral folds, column foot half as long as column . . . . go to 19

19a. Basal (posterior) callus is an erect warty knob on a stalk, flower yellow, lip with external brown pustules along the margins of lateral lobes . . . . *S. pustulosa*

19b. Basal callus just a mound without a stalk . . . . go to 20

20a. Column with definite wings, flower white, interior lip with purple spots. . . . . *S. pastorellii*

20b. Column not winged, flower green with red spots on lip margin . . . . *S. luerorum*

Stenia angustilabia D. E. Bennett & Christenson, *Lindleyana* 13(2): 88. 1998. Type: Peru, probably Department of Cuzco, *Bennett 7220* (holotype: CUZ).

DESCRIPTION The flowers are pale cream-white to green-yellow. The sepals and petals do not spread fully open. The rigid three-lobed lip is concave; the lateral lobes are shallow, broadly rounded and erect, the long thin central lobe is shaped like a boat with a long bow and has an acute apex. The callus is narrow, arising from a thick hirsute swelling at the base of the lip, continuing apically as three keels which terminate in a raised bidentate, arcuate tip with a median keel. The column is nearly straight and held at right angles to the column foot, the rostellum is ligulate, and the column base and foot are pubescent. PLATE 121.

MEASUREMENTS Leaves 9 cm long, 2.5 cm wide; inflorescence 7 cm long, with two bracts; dorsal sepal 2.4 cm long, 1.2 cm wide; lateral sepals 2.6 cm long, 1 cm wide; petals 2.2 cm long, ca. 1.15 cm wide; lip 2.2 cm long, 1.3 cm wide; column 1.5 cm long, 0.4 cm wide, column foot 0.5 cm long.

ETYMOLOGY Latin *angustus*, narrowed, and *labia*, lip, referring to the narrow apical lip.

DISTRIBUTION AND HABITAT Known from Ecuador to southern Peru.

FLOWERING TIME May to June in cultivation in Lima.

Stenia aurorae D. E. Bennett & Christenson, *Lindleyana* 13: 91. 1998. Type: Peru, Junín, Tarma, between Carapata and Huasahuasi, 2200 m, 2 August 1994, *O. del Castillo ex Bennett 6640* (holotype: NY).

DESCRIPTION The sepals and petals are pale yellow-green. The lip is pale buff-yellow with the lateral lobes of the lip having an intense mahogany-brown edge, the inner surface of the lobes has pale red-brown spots, and the central and transverse calluses have faintly marked spots. The three-lobed lip is saccate and rigid, the lateral lobes are erect, the midlobe is triangular with involute margins forming a tubular apex, and the lip is rigidly joined to the foot by a short broad claw. The disc of the lip internally has 13 low slightly raised ribs or keels on the basal half. The central keels are thicker. The keels terminate into the erect callus which runs transversely across the middle lobe of the lip at mid length, producing a wavy knobby wall midway transversely on the inside lip. The column is stout, dorsally with two lateral furrows with marginal ridges, and held at right angles to the column foot. The base and foot of the column are sparsely pubescent. The column foot is 6 mm long. The stipe has 2 or 3 minute retrorse teeth on each side. The viscidium is rhombic. PLATE 122.

MEASUREMENTS Leaves 15–22 cm long, 2–3 cm wide; inflorescence 5 cm long with one bract; dorsal sepal 2 cm long, 1.3 cm wide; lateral sepals 2.2 cm long, 1.4 cm wide; petals 1.8 cm long, 1.2 cm wide; lip 1.2 cm long, 2.3 cm wide; column 1.2 cm long, column foot 0.7 cm long.

ETYMOLOGY Latin *auratus*, flecked with gold, in reference to the flower color.

DISTRIBUTION AND HABITAT Known only from Peru, at 2200 m, in wet montane forest.

FLOWERING TIME July to September.

Stenia bismarckii Dodson & D. E. Bennett, *Icon. Pl. Trop.*, ser. 2, 2: plate 179. 1989. Type: Peru, Pasco, Oxapampa, 1850 m, 20 May 1987, *Bennett & A. Vargas 3831* (holotype: MO).

DESCRIPTION The sepals and petals are pale blue-green or white, the lip is white or yellow mottled with pale purple or red spots, and the column is white. The three-lobed lip is concave with erect lateral lobes, the lip base is saccate, and the midlobe is retuse. The claw of the lip is sharply recurvate and articulated to a sharply sigmoid shape with the narrow column foot. The callus is a free lobulate plate that transverses the midlobe without ribs or keels. The column has two obtuse wings and hoods over the ligulate rostellum. The column foot is longer than the column length and held at right angles to the column. The apical column extends frontally more than the lip apex. The stipe has basally fimbriate margins. PLATE 123.

COMMENT Plate 912 in *Icones Plantarum Tropicarum* is labeled "*Chondrorhyncha stenioides*" but actually shows *Stenia bismarckii*.

MEASUREMENTS Leaves 17 cm long, 3 cm wide; inflorescence 4 cm long with one bract; dorsal sepal 2.1 cm long, 1.3 cm wide; lateral sepals 2.3 cm long, 1.1 cm wide; lip 1 cm long, 2 cm wide (when flattened); column 1.4 cm long, 0.4 cm wide, column foot 0.8 cm long.

ETYMOLOGY Named after Klaus von Bismarck, the first collector of the species.

DISTRIBUTION AND HABITAT Known from

Ecuador and Peru, at 1600–1850 m elevation, in cool wet forest.

FLOWERING TIME January to May.

Stenia bohnkiana V. P. Castro & G. F. Carr, *Orch. Digest* 69: 49. 2005. Type: Brazil, Bahia, Buerarema, 800 m, *Castro Neto 08* (holotype: SP).

DESCRIPTION The flowers are clear pale yellow and the lip is dotted with purple on the interior and uppermost parts. The lip is saccate, slightly trilobed, and fleshy. The lateral lobes are sub-auricular and upright at the uppermost border, not folded out, the center lobe is pointed, and the apical lip is closed along the edges so that there is no apical opening. The callus has five or six parallel keels that terminate in the middle lobe. The column has obtuse wings and is at right angles to the column foot. PLATE 124.

COMMENT *Stenia bohnkiana* at first glance looks like *S. pallida* and has been considered to be that species for years. *Stenia bohnkiana* differs by having sepals and petals that are more acuminate, upright lateral edges to the lip which do not roll outward, an acute point to the apical end of the lip, and no opening to the lip apically, among other differences.

MEASUREMENTS Leaves 9.0–11.5 cm long, 1.7–2.0 cm wide; inflorescence 2–3 cm long with one bract; dorsal sepal 2.8 cm long, 1.2 cm wide; lateral sepals 3.9 cm long, 1.5 cm wide; petals 2.6 cm long, 1 cm wide; lip 2 cm long, 0.8 cm wide; column 1.3 cm long, 0.5 cm long, column foot 1.3 cm long.

ETYMOLOGY Named for Erwin Bohnke, a Brazilian orchidologist.

DISTRIBUTION AND HABITAT Known from the Atlantic rain forest of Bahia, Brazil, 30–40 km from the Atlantic Ocean, at 800 m in elevation.

FLOWERING TIME February to March.

Stenia calceolaris (Garay) Dodson & D. E. Bennett, *Icon. Pl. Trop.*, ser. 2, 2: plate 180. 1989. Basionym: *Chaubardiella calceolaris* Garay, *Orquideología* 4: 148. 1969. Type: Peru, Amazonas, *Hutchison & Wright 6815* (holotype: UC).

SYNONYM
Chaubardiella parsonii hort. ex Jenny, *Orchidee (Hamburg)* 40: 200. 1989. *nom. nud.*

DESCRIPTION The sepals are pale green or green-yellow with scattered rose-red spots, the petals are green-yellow densely spotted with rose, the lip is cream-yellow to yellow with red spots, the callus teeth are white, and the column is green-white to yellow with pale red markings. The sepals and petals are concave, giving the flower a cupped appearance. The lip is deeply concave, forming a broad pouch. The disc has an oblong, raised callus, which is free at its tridentate apex, and has two or three pairs of teeth that arise from the pouch like sides which project upward and inward to the center. The column has two very obtuse wings and is held at right angles to the column foot. PLATE 125.

MEASUREMENTS Leaves 10 cm long, 2 cm wide; inflorescence 3–4 cm long with one bract; dorsal sepal 1.8 cm long, 1 cm wide; lateral sepals 2.4 cm long, 1.2 cm wide; petals 1.8 cm long, 1.1 cm wide; lip 1 cm long, 0.7 cm wide; column 1 cm long, column foot 0.5 cm long.

ETYMOLOGY Latin *calceolatus*, slipper-shaped, referring to the shape of the lip.

DISTRIBUTION AND HABITAT Known from Ecuador and Peru, at 2000–2100 m elevation, in cool wet forests.

FLOWERING TIME November to February.

Stenia christensonii D. E. Bennett, *Brittonia* 50(2): 189. 1998. Type: Peru, Pasco, Oxapampa, Puerto Bermúdez, Sector Milagros along Pichis River, 280 m, 10 November 1991, *O. del Castillo ex Bennett 5387* (holotype: NY; isotype: MOL).

DESCRIPTION The sepals, petals, and column are a pale clear green, the lip is pale yellow-green with scattered pale brown spots on the exterior of the lateral lobes, the interior markings are pale red-brown, and the anther is pale cream-yellow. The flowers are bell-shaped with overlapping segments. The sepals and petals are rounded at the apex. The three-lobed lip is saccate, attached to the column foot by an elongate papillose basal claw that is sharply recurved and conduplicate in the middle and held in close proximity to nearly touching the rear surface of the callus. The lateral lip lobes are erect with an acute transverse fold near the mid length of the lip, the leading margin of the lateral lobe is undulate. The midlobe of the lip is broadly triangular and conduplicate. The callus is biseriate, the posterior callus is erect, high, three-tiered, papillose, and extends along the midvein as a low thick rib or keel; the anterior callus is a transverse lamella, and the inner surface is collicuate. The column is obscurely winged, with scattered short trichomes over its surface, and held at right angles to the column foot. The apical column extends frontally more than the lip apex.

MEASUREMENTS Leaves 14.5 cm long, 3.7 cm wide; inflorescence 6 cm long with one bract; sepals 1.7–1.9 cm long, 1.4 cm wide, lateral sepals longer; petals 1.6 cm long, 1.5 cm wide; lip 0.75 cm long, 0.7 cm wide (folded); column 1.2 cm long, 0.7 cm wide, column foot 0.7 cm long.

ETYMOLOGY Named for Eric Christenson, a well-known orchid taxonomist.

DISTRIBUTION AND HABITAT Known only from Peru, at 280–1830 m in elevation, in tropical and cool wet forest.

FLOWERING TIME Late August.

Stenia glatzii Neudecker & G. Gerlach, *Orquideología* 21(3): 259. 2000. Type: Ecuador, near Gualaquiza, 800 m, collected by Antón Glatz, 15 April 1979 (holotype: M).

DESCRIPTION The sepals and petals are beige and the lip is yellow with fine red dots in the basal area and lateral lobes. The lip is three-lobed, the lateral lobes are broadly oval, and the anterior margins of the lateral lip lobes have small teeth or points on the margins. The lip middle lobe is concave and acute at its apex. The callus is broad and basal, consisting of 15 longitudinal keels ending apically in slightly raised teeth. The column is slightly winged. The column foot is held at about 130 degrees from the column. The stipe is short.

COMMENT *Stenia glatzii* seems to be an intermediate between *S. pallida* and *S. saccata*, if not evolving from a hybrid of the two.

MEASUREMENTS Leaves 10 cm long, 3 cm wide; inflorescence 5 cm long with one bract; dorsal sepal and petals 3 cm long, 1.6 cm wide; lateral sepals 3.5 cm long, 1.6 cm wide; lip 2 cm long, 1.25 cm wide; column 1 cm long, 0.6 cm wide, column foot 0.6 cm long.

ETYMOLOGY Named for Antón Glatz, who collected the original plant.

DISTRIBUTION AND HABITAT Known only from Ecuador, at 800 m elevation.

FLOWERING TIME April.

Stenia guttata Reichenbach f., *Gard. Chron.* n.s., 14: 134. 1880.

SYNONYM
Chondrorhyncha guttata (Reichenbach f.) Garay, *Bot. Mus. Leafl.* 26: 27. 1978. Type: Peru, *Davis s.n.* (holotype: W).

DESCRIPTION The sepals are pale green and the sepals have a few purple spots. The petals are of similar coloration but more densely spotted, the lip is yellow with larger purple spots on the lateral lobes, the callus teeth are cream-white, and the column is pale green with a yellow apex. The lip is lightly three-lobed, with the lateral lobes bending outward, but not curling or folding. The callus is transverse and seven-toothed, the points angled apically. The column apex does not extend forward farther than the lip apex. The column foot is held at approximately right angles to the column. PLATE 126.

MEASUREMENTS Leaves 10–14 cm long, 2–4 cm wide; inflorescence 6 cm long with one bract; dorsal sepal 2.5 cm long, 1 cm wide; lateral sepals 2.4 cm long, 1.3 cm wide; petals 2.1 cm long, 1.1 cm wide; lip 2 cm long, 0.5 cm wide (folded); column 2 cm long, column foot 0.5 cm long.

ETYMOLOGY Latin *guttatus*, spotted, referring to the flowers.

DISTRIBUTION AND HABITAT Known only from Peru, at 1900 m elevation, in montane wet forest on small trees.

FLOWERING TIME April to July.

Stenia jarae D. E. Bennett, *Lindleyana* 7: 80. 1992. Type: Peru, Huánuco, Leoncio Prado, Las Palmas, 1000 m, 20 December 1990, *E. Jara P. s.n. ex Bennett 4711* (holotype: USM).

DESCRIPTION The dorsal sepal is semitranslucent, very pale green-yellow with 11 to 13 indistinct pale green veins, the lateral sepals are pale cream-yellow, and the petals are very pale green-yellow. The lip is opaque cream-yellow with pale chestnut-brown spots on its interior surface, the column is pale green to green-white, the anther is white, and the column foot has longitudinally aligned dark brown or purple-red short streaks and spots. The long-clawed, three-lobed lip is basally saccate, conduplicate, and tapered toward the apex. A thick midvein extends from near the base of the lip becoming an elevated carinate ridge at the semiobovate apex, dividing the transverse callus. The callus has three teeth on each side, and the apical callus margins are simple, blunt, and irregular. The column is abruptly arcuate apically at the stigma, prominently winged near or below the mid column length, and narrowing below to the column foot. PLATE 127.

COMMENT Both *Stenia jarae* and *S. lillianae* have a column that is bent or curved at the stigma of the column apex and a column foot that is held at about 160 degrees to the column, whereas other species of *Stenia*, excepting *S. glatzii* and *S. saccata*, hold their column and column foot at right angles. *Stenia lillianae* has larger flowers than *S. jarae*. The two species also differ in their column wings: those on *S. jarae* are acute, those on *S. lillianae* obtuse.

MEASUREMENTS Leaves 12 cm long, 5 cm wide; inflorescence 6–7 cm long with one bract; dorsal sepal 1.8 cm long, ca. 0.85 cm wide; lateral sepals 2.1 cm long, 1 cm wide; petals 2 cm long, 0.8 cm wide; lip 1.65 cm long, 0.8 cm wide; column 1.5 cm long, 0.4 cm wide, column foot 0.8 cm long.

ETYMOLOGY Named for E. Jara P., who helped in the collection of the type species.

DISTRIBUTION AND HABITAT Known only from Peru, at 1000–2200 m in elevation, in wet montane forest.

FLOWERING TIME Sporadically throughout the year.

Stenia lillianae

Stenia lillianae Jenny ex D. E. Bennett & Christenson, *Brittonia* 46(1): 46. 1994. Type: Peru, San Martín, Moyobamba, 4 October 1992, *D. E. Bennett 5684* (holotype: NY).

DESCRIPTION The flowers are semitranslucent pale green-yellow, the lip is cream-yellow externally and internally with red-brown spots, and the column is pale green with red-brown spots on the column foot. The three-lobed lip is saccate and extremely fleshy. The lip claw is long. The lateral lobes of the lip are pinched together; the midlobe is narrowly triangular, acute, with a high elongate keel along its median. The callus is an irregularly margined transverse ridge with two longer unequal teeth, one on each side; it is divided by a high rounded keel that is shorter than the callus. The column is strongly arcuate apically with large broad wings, and

the column foot has a papillose base. The long stipe has small pectinate appendages (fimbriate margins) at its apex. The small viscidium is quadrate. PLATE 128.

COMMENT For a comparison of *Stenia lillianae* and *S. jarae*, see *S. jarae*.

MEASUREMENTS Leaves 6.5–17.5 cm long, 3.2–4.3 cm wide; inflorescence 3 cm long with one bract; dorsal sepal 2.5 cm long, 1.5 cm wide; lateral sepals 3.1 cm long, 1.5 cm wide; petals 2.4 cm long, 1.35 cm wide; lip 2.2 cm long, 1 cm wide (in natural position); column 1.3 cm long, column foot 1 cm long.

ETYMOLOGY Named for Lil Severns, an orchid grower. See discussion under *Stenia pustulosa*.

DISTRIBUTION AND HABITAT Known only from Peru, at 1600 m elevation, in deep shade in wet montane forest.

FLOWERING TIME Several times a year as new growths mature.

Stenia luerorum

Stenia luerorum D. E. Bennett & Christenson, *Lindleyana* 13(2): 91. 1998, as "*lueriorum*." Type: Peru, Pasco, Oxapampa, Chontabamba, 1600 m, 17 April 1994, *O. del Castillo ex Bennett 6534* (holotype: NY).

DESCRIPTION The flowers are a pale translucent green, the lip is cream-yellow, and the lateral lip lobes are green-yellow with dark garnet-red spots. The column is white tinted with pale green, the column base and foot have short vertically aligned dark purple-red bars and spots, the anther is wax-white, and the pollinia are ivory white. The flowers are spreading (star-shaped, not bell-shaped). The three-lobed lip is proportionately small for the genus. The lateral lip lobes are erect having revolute leading edges. The midlobe is conduplicate, with involute overlapping margins that form the tubular apex. The callus has two parts, the basal callus has an erect high multiridged keel with scattered papillose clusters while the apical callus is a transverse irregular apical ridge. The column is semiterete with a 7-mm-long foot held at a right angle to the column. The foot is papillose on the ventral surface. The apical column extends frontally more than the lip apex. The stipe is linear with denticulate margins. PLATE 129.

MEASUREMENTS Leaves 2.5–6.5 cm long, 0.9–1.8 cm wide; inflorescence 5 cm long with one bract; sepals and petals 2.4 cm long, 1.2 cm wide; lip 0.6 cm long, 1.7 cm wide (unfolded); column 1.25 cm long, column foot 0.75 cm long.

ETYMOLOGY Named to honor Dr. and Mrs. Carlyle Luer.

DISTRIBUTION AND HABITAT Known only to Peru, at 1600 m elevation, in wet montane forest.

FLOWERING TIME September and October and again in April.

Stenia nataliana R. Vásquez, Nowicki & R. Müller, *Rev. Soc. Boliv. Bot.* 3(1–2): 33. 2001. Type: Bolivia, La Paz, B. Saavedra, between Charazani and Camata, 2600 m, *C. Norwicki & R. Müller 2173* (holotype: LPB; isotype: VASQ).

DESCRIPTION The sepals and petals are yellow with red spots at the base; the lip has many red spots on its exterior and often the red on the sides of the lip lies in the same lines as the interior callus. The three-lobed lip is saccate. The lateral lobes are erect and have a glandular interior. The lip is held so that the apical point is the lowest point of the lip. The basal callus consists of 10 keels with toothed ends apically. The column is slightly bent and the column foot is pubescent.

COMMENT *Stenia nataliana* is very similar to *S. vasquezii*, and indeed might be better thought of as a form of *S. vasquezii*, the differences being the size of the petals and sepals, a slight variation in the lip margin, how the lip is held, and the color of flower. The callus keels of *S. nataliana* are drawn all the same length, whereas those of *S. vasquezii* are longest medially becoming shortened laterally.

MEASUREMENTS Leaves 15 cm long, 2 cm wide; inflorescence 5 cm long with one bract; dorsal sepal 2.2 cm long, 1.7 cm wide; lateral sepals 2.5 cm long, 1.4 cm wide; petals 2.1 cm long, 1.3 cm wide; lip 1.7 cm long, 1 cm wide; column 0.8 cm long, column foot 0.4 cm long.

ETYMOLOGY Named for Natalia Dakers, in recognition for work in botanical investigation in Bolivia.

DISTRIBUTION AND HABITAT Known only from Bolivia and possibly Peru, at 2600 m in elevation, growing on limbs in humid cloud forests.

FLOWERING TIME November.

Stenia pallida Lindley, *Bot. Reg.* 23: plate 1991. 1837. Type: Guyana, Demerara, *Barker s.n.* (holotype: K); type species of *Stenia*.

DESCRIPTION The sepals and petals are pale cream-yellow. The lip is yellow on the outside, the interior of the lateral lobes is intense yellow with red spots, the midlobe interior is paler yellow, the keels of the disc are cream-white faintly spotted with pale rose, the column is pale green-white faintly spotted with pale rose, the anther is white with small faint rose spots, the viscidium is pale brown-purple, and the pollinia are dark yellow. The lip is deeply concave, indistinctly three-lobed, with revolute lateral lobes. The callus is a disc transversed by 12–13 apically toothed longitudinal keels. The column is short and straight with obtuse wings, and the column foot is held at a right angle to the column. The pollinia are attached to a flat orbicular retuse viscidium by a short broad stipe. PLATE 130.

MEASUREMENTS Leaves 15 cm long, 4 cm wide; inflorescence 5 cm with one bract; dorsal sepal 3 cm long, 1.5 cm wide; lateral sepals 3.5 cm long, 1.6 cm wide; petals 3.2 cm long, 1.4 cm wide; lip 2.6 cm long, 1.2 cm folded; column 1.2 cm long, 0.5 cm wide, column foot 0.7 cm long.

ETYMOLOGY Latin *pallidus*, pale, referring to the color of the flower.

DISTRIBUTION AND HABITAT Known from Bolivia, northern Brazil, Colombia, Ecuador, Guyana, Peru, Venezuela, and Trinidad, at 100–2000 m elevation, in montane wet forest.

FLOWERING TIME April to June.

Stenia pastorellii D. E. Bennett, *Lindleyana* 7(2): 83. 1992. Type: Peru, San Martín, Moyobamba, near Jepelacio, 950 m, 20 December 1989, *R. Villena ex Bennett 4879* (holotype: USM).

DESCRIPTION The sepals are white, the petals are translucent white, the lip exterior is glossy white with a cream-colored apex, the interior is marked with spots and short purple streaks, the column is white, and the pollinia are yellow. The lip is three-lobed, deeply and rigidly saccate, with a narrow claw which is flanked by a transverse callus. The callus is two-parted having a rather elongate and complex median basal callus with many irregular knobs and folds and a transverse apical callus with four tubercles on each side. The column is clavate, straight, with triangular wings, and the column foot is rather long tapering to a sharply recurved toothed tip with papillose lower markings. The column foot is held at right angles to the column. The apical column extends frontally more than the lip apex. The stipe has fimbriate margins. PLATE 131.

MEASUREMENTS Leaves 14 cm long, 2.5 cm wide; inflorescence 7 cm long with one bract; dorsal sepal 2.2 cm long, 1 cm wide; lateral sepals and petals 2 cm long, 1 cm wide; lip 1 cm long, 0.7 cm wide (from side); column 1.2 cm long, column foot 0.7 cm long.

ETYMOLOGY Named for Pastorell.

DISTRIBUTION AND HABITAT Known only from Peru, at 950 m.

FLOWERING TIME Sporadically throughout the year.

Stenia pustulosa D. E. Bennett & Christenson, *Brittonia* 46(1): 48. 1994. Type: Peru, Junín, Tarma, southeast of Contayapaccha, Quebrada Seca, 1800 m, 22 April 1992, *O. del Castillo ex Bennett 5552* (holotype: NY).

SYNONYM
Stenia lillianae Jenny, *Orchidee (Hamburg)* 40(3): 199. 1989, *nom. nud.*

DESCRIPTION The translucent flowers are pale yellow. The lip is pale green externally with pustules along the apical earlike margins of the lateral lobes, which are pale brown turning purple apically. The interior of the lip is streaked and spotted brown-red; the midlobe is very pale yellow with a few brown-red spots along the margins. The callus is pale white with green-yellow teeth tipped brown-red. The column is pale yellow-white and the column foot has faint lavender streaks and spots. The anther is pale cream-white, and the pollinia are shiny pale yellow. The petals are minutely apiculate. The three-lobed lip has erect lateral lobes, and the midlobe is concave to the point the lateral edges touch to form a beak. The callus is in two parts; the basal callus is an erect sparsely papillose knob or point, and the apical callus is a transverse ridge of 10–12 teeth extending transversely for nearly the entire width of the lip. The column is lightly arcuate with barely discernible wings, and the column foot is prominent and pubescent and held at a right angle to the column. The apical column extends frontally more than the lip apex. The stipe is serrulate. PLATE 132.

COMMENT Rudolf Jenny was very helpful in providing from his collection the flower of the type specimen on which he based his original description of *Stenia lillianae* Jenny. This flower is an exact match for *S. pustulosa*. Jenny was correct that the species was new at the time, but the rules of publication made his name invalid. In the original publication Jenny forgot to publish the herbarium where the type specimen would be housed, making the name invalid. Eric Christenson, in trying to validate the name *"lillianae,"* named another species that was also grown by Lil Severns as *S. lillianae* Jenny ex D. E. Bennett & Christenson, and named the species that Jenny described in his 1989 article *S. pustulosa*.

MEASUREMENTS Leaves 6.5–13.0 cm long, 2.0–2.3 cm wide; inflorescence 8 cm long with one bract; sepals 2.6–2.7 cm long, 1.1–1.2 cm wide; petals 2.4 cm long, 1.4 cm wide; lip 0.7 cm long and wide (in natural position); column 1.4 cm long, column foot 0.7 cm long.

ETYMOLOGY Latin *pustulosus*, having pustules, referring to the blisterlike lesions on the lateral lobes of the lip.

DISTRIBUTION AND HABITAT Known only from Peru, at 1600–1890 m in elevation, in cool wet montane forest and lower levels of cloud forest.

FLOWERING TIME March to April and October to November as new growth matures.

Stenia saccata Garay, *Orch. Rev.* 77: 152. 1969.
Type: Ecuador, along Río Zamora, collected by J. Strobel, cultivated by Henry Scarfield Jr., 12 November 1968 (holotype: AMES), type of the genus *Dodsonia*.

SYNONYMS
Chaubardiella saccata (Garay) Garay, *Orquideología* 4: 148. 1969.
Dodsonia saccata (Garay) J. D. Ackerman, *Selbyana* 5: 119. 1979.
Dodsonia falcata J. D. Ackerman, *Selbyana* 5: 118. 1979.
Stenia falcata (J. D. Ackerman) Dressler, *Lankesteriana* 5(2): 93. 2005.

DESCRIPTION The flower is white, cream, or yellow; the lip has red spots. The basal portion of the lip is saclike and very fleshy. The apex of the lip is bilobed; the lobes are acute, fimbriate on the interior margins, curling at the apex and clasping the column. The basal callus is transverse and platelike, with an irregular margin and lateral thickenings. The apical callus is bilobed, fleshy, with its apical margin lacer-

ated. The column is straight; the column foot is small and the angle between the column and the column foot is almost 180 degrees. There are four very elongate and cylindrical pollinia, a shorter combined with a longer one in pairs. The stipe is minimal, inserting on the edged of the heart-shaped viscidium. PLATE 133.

COMMENT The basal callus of *Stenia saccata* has a more dentate edge, especially on the lateral portions, whereas the drawing of *S. falcata* shows the same plate but without the teeth. Other than this there is no difference between the two species. Both were collected near Zamora in southeastern Ecuador, but over time their habitat has vanished. They were assumed to be never collected again, and some individuals have speculated that the two species were extinct. In the late 1990s, plants were discovered in cultivation in a German collection. These plants of *Stenia* were collected near Gualaquiza with the same habitat conditions, but at a lower elevation. This allowed live material to be compared with the earlier herbarium specimens. Neudecker and Gerlach (2005: 258) presented the argument that the two species are actually one species with some variability. I have followed them here in this description and classification and combined the two, though some will argue they are separate species.

MEASUREMENTS Leaves 8 cm long, 2 cm wide; inflorescence 2.5 cm long with one bract; dorsal sepal 2 cm long, 1.5 cm wide; lateral sepals 2.5 cm long, 1.3 cm wide; petals 2.2 cm long, 1.4 wide; lip 1.7 cm long; column 0.5 cm long, column foot 0.5 cm long.

ETYMOLOGY Latin *saccatus*, pouched or bag-shaped, referring to the shape of the lip.

DISTRIBUTION AND HABITAT Known only from Ecuador, at 800 m in elevation.

FLOWERING TIME Not recorded.

Stenia stenioides (Garay) Dodson & R. Escobar, *Orquideología* 18(3): 206. 1993. Basionym: *Chondrorhyncha stenioides* Garay, *Bot. Mus. Leafl.* 26(1): 26, plate 1978. Type: Ecuador, Pastaza, *Stacy s.n.* (holotype: AMES).

DESCRIPTION The flowers are white to green-white to yellow, the lip interior has obscure purple spots and a brown to maroon mottling on the lip exterior, and the column foot often has longitudinal maroon streaking. The dorsal sepal and petals are mildly concave and the flower does not spread fully open. The lateral sepals are only basally concave. The complexly trilobed lip has a claw that is not recurvate. The lip midlobe is concave to the point the lateral edges touch to form a beak. The callus is two-parted, the basal portion is sigmoid in a longitudinal direction and fleshy, and the apical callus is toothed apically. The column is slightly bent with minimal wings and extends frontally to only half the length of the lip. The column foot ventral surface is slightly longer than the column ventral surface, and the column foot is held at a right angle to the column. PLATE 134.

COMMENT Plate 912 in *Icones Plantarum Tropicarum* is labeled "*Chondrorhyncha stenioides*" but actually shows *Stenia bismarckii*. The plant labeled "*Stenia stenioides*" in *Brittonia* (46: 50) is most likely *S. bismarckii*. The illustration of *S. stenioides* in *Native Ecuadorian Orchids* (Dodson and R. Escobar 1993, 5: 1023) shows a white flower with a yellow column and no other markings. The plant in Plate 134 of the present volume has a green-yellow flower with brown. The original description says the flower is white or off-white with obscure purple spots. This is color variation within the species.

MEASUREMENTS Leaves 13 cm long, 3 cm wide; inflorescence 3 cm long with one bract; dorsal sepal 2 cm long, 1.6 cm wide; lateral sepals 2.4 cm long, 1 cm wide; petals 2 cm long, 1.2 cm wide; lip 1.3 cm long, 1 cm wide; column 0.9 cm long, column foot 0.9–1.0 cm long.

ETYMOLOGY Named for the genus *Stenia* and the Latin *oides*, appears like; when first described, the species was classified with *Chondrorhyncha* but the author realized it looked like a *Stenia* species.

DISTRIBUTION AND HABITAT Known only to Ecuador.

FLOWERING TIME Not recorded.

Stenia uribei P. Ortíz, *Orquideología* 23(1): 29. 2004. Type: Colombia, Nariño, cultivated by Carlos Uribe Vélez, Bogotá, August 2003, *P. Ortíz 1142* (holotype: HPUJ).

DESCRIPTION The flower is pale yellow with small purple spots inside the base of the lip and on the column foot. The three-lobed lip is boat-shaped having raised lateral lobes and an apical lobe that is long and tubular. The callus is straight, consisting of five raised keels terminating in a toothed, free transverse plate, the two most lateral keels are the shortest while the middle keel is slightly shorter than the longest keels. The column is broad and straight and the column foot is held at about 110 degrees from the column. PLATE 135.

MEASUREMENTS Leaves 12 cm long, 2.1 cm wide; inflorescence 7 cm long with one bract; dorsal sepal 2.1 cm long, 1.3 cm wide; lateral sepals and petals 2 cm long, 1 cm wide; lip 2.5 cm long, 1.2 cm wide; column 1.4 cm long, column foot 0.5 cm long.

ETYMOLOGY Named for Carlos Uribe Vélez of Bogotá, who grew the species.

DISTRIBUTION AND HABITAT Known only from southern Colombia in the Department of Nariño.

FLOWERING TIME August.

Stenia vasquezii Dodson, *Icon. Pl. Trop.*, ser. 2, 4: plate 384. 1989. Type: Bolivia, Cochabamba, Chapare, 1800 m, 25 December 1979, *R. Vásquez 231* (holotype: SEL).

DESCRIPTION The flowers are white or green-yellow; the sepals and petals have brown-red spots arranged somewhat in lines with the spots extending to the apexes of the sepals and petals. The lip margin is red, the lateral margins of the interior lip have red spots arranged in lines, and the column is green-yellow with red longitudinal lines on the ventral base. The sepals and petals are spreading. The obscurely three-lobed lip is saccate, the claw is elongate, straight, and pulverulent. The lip is triangular when viewed from above, the internal lateral margins are minimally pubescent, and the outside of the apex is sparsely pubescent. The lip upper margin is held horizontal so that the apex is not the lowest point in the lip. The callus is a series of 10–11 thickened longitudinal keels arranged in a transverse ledge. The column is stout, wingless, and pubescent especially on the column foot. The stigma extends laterally to the margins of the lateral sides of the column, so that it appears to be a cut transversely across the column. PLATE 136.

COMMENT *Stenia vasquezii* is very similar to *S. nataliana* but smaller. For details, see discussion under *S. nataliana*.

MEASUREMENTS Leaves 11 cm long, 2.5 cm wide; inflorescence 5.5–7.0 cm long with one bract; dorsal sepal and petals 1.7 cm long, 1.1 cm wide; lateral sepals 1.9 cm long, 1 cm wide; lip 1.6 cm long, 1.2 cm wide; column 1 cm long, column foot 0.4 cm long.

ETYMOLOGY Named for Roberto Vásquez, who discovered the species.

DISTRIBUTION AND HABITAT Endemic to Bolivia and Peru, at 1800 m in elevation.

FLOWERING TIME October.

Stenia wendiae D. E. Bennett & Christenson, *Brittonia* 46: 51. 1994. Type: Peru, Junín, Chanchamayo, Sector Chipes of the San Vicente mine, 1800 m, 10 December 1992, *O. del Castillo ex D. Bennett 5968* (holotype: NY).

DESCRIPTION The flowers are pale, semitranslucent green; the lip is white tinged green towards the base with dark purple spots on the concave center. The flowers are spreading. The sepals are subequal and apiculate with both the sepals and petals nearly oval. The three-lobed lip has a wedge-shaped base. The column foot and claw fold into a high arch with a distinctive pointed tip. The lateral lip lobes are erect and broadly rounded, and the midlobe is much smaller and acute. The callus is two-parted, the basal callus is flattened, the transverse callus is carinate and terminates in a prominent free tooth toward the midline, which is continuous with a central low transverse ridge across the middle of the lip. The column is lightly arcuate, concave, and papillose basally, with a thick, recurved column foot held at a right angle to the column. The column apex extends slightly more forward than the apex of the lip. The stipe is thickened with fimbriate (described as cleft) margins.

MEASUREMENTS Leaves 6–9 cm long, 2.0–2.9 cm wide; inflorescence 5 cm long with one bract; sepals 1.5 cm long, 1 cm wide; petals 1.5 cm long, 1.2 cm wide; lip 0.5 cm long, 0.6 cm wide (when expanded); column 1.2 cm, column foot 0.5 cm long.

ETYMOLOGY Named for Wendy de von Bismarck, who traveled with the describing author.

DISTRIBUTION AND HABITAT Known only to Peru, at 1800 m in elevation, in cool wet montane forest in dense shade.

FLOWERING TIME February to April.

• •

Dodsonia J. D. Ackerman, *Selbyana* 5: 118. 1979.

Dodsonia was established in 1979 by James Ackerman and consisted of the one (or two) species listed here as *Stenia saccata* (and *S. falcata*). Ackerman felt that these species were distinct enough from *Stenia* to justify their own genus; however, molecular analysis does not support or encourage their separation from *Stenia*. *Stenia glatzii* is considered an intermediate species between the two genera. The molecular data determine that *S. glatzii* is imbedded in *Stenia*, making it difficult to separate the two genera, so Dressler (Whitten et al. 2005) combined the two genera.

The plants of *Dodsonia* have no pseudobulbs, the foliage has a distinct sheath and appears like a fan, and the leaves are narrow and clearly veined. The inflorescence is one-flowered, emerging from the base of the foliage. The lip is saclike and the lip lobes are narrow, folded, and incurved, with a transverse ridge on the inferior surface. The lateral lip border has sharp lobules (deeply notched). There are two sets of calluses, one on the lateral walls appearing as warty protuberances and another at the lip apex that is flat and bluntly toothed apically. The column has no keels on the ventral surface. The four flat pollinia are imposed with a short stipe and a flat viscidium. Though initially thought to be two species, when Neudecker and Gerlach (2000) examined the calluses of a recently found live plant, they found variability among the calluses, combining the features of both previously described species. They concluded that there is so far only one species in the genus *Dodsonia*.

Plants of this genus are represented by two specimens collected in the town of Zamora, Ecuador, in an area where all the native vegetation has been disrupted, lending to the belief that these orchids are extinct in nature. They were found on the trunks of trees in dense humid forest at 700–1500 meters in elevation. Ecuagenera has recollected this (these) species recently and is working on propagating it (them).

Etymology
Dedicated to Calaway H. Dodson, a prominent orchid taxonomist.

Stenotyla Dressler, *Lankesteriana* 5(2): 96. 2005.

Robert Dressler described the genus *Stenotyla* in 2005 for a group of species closely related to *Cochleanthes*. Based on molecular data and morphological similarities, these species could not remain in *Chondrorhyncha*. The earliest described species, *S. picta*, has been transferred to several genera over the years. Originally it was assigned to *Warczewiczella*, then *Cochleanthes*, then *Chondrorhyncha*. The narrowing of the parameters for inclusion in *Chondrorhyncha sensu stricto* left this small group of species without a genus, hence the formation of *Stenotyla*.

The light to dark green *Stenotyla* leaves are arranged in a fan shape, with each new leaf of a growth alternating sides. There is a very short pseudobulb, more like a stem, which is palpable through the leaf sheath. The leaves are strap-shaped, becoming widest at the apical three-quarters of the leaf length, with an acute apex. Basally the leaf tapers to the junction of the leaf sheath. The flowers are somewhat erect in presentation but remain under the foliage. They generally have reflexed or spreading lateral sepals. The tubular lip lateral margins clasp the column. The narrow nearly flat callus is medial and basal, not extending laterally onto the lip sides. The callus is two- to four-toothed, the teeth minimally free at their apex and adpressed to the lip surface, and there is no apical thickening of the lip. The claw of the lip (the part of lip that attaches to the column foot before it widens out to the blade of the lip) is short or minimal. The column is only slightly curved with the clinandrium angled back below the apex of the column. The column is widest at the stigma but not clublike. Instead of distinct wings the column has a margin that creates a thin plane of tissue on the sides of the column. The short column foot angles forward slightly and is narrower than the column. The stipe is small and the viscidium is shaped like a long triangle.

The flowers appear similar to those of *Ixyophora*, but molecular data place the two genera apart. The genus *Stenotyla* is more closely related to *Cochleanthes* than to any other genus in the *Huntleya* group, based on molecular data.

Etymology

Greek *stenos*, narrow, and *tylo*, callus, describing the shape of the lip callus.

List of the Species of *Stenotyla*

Stenotyla estrellensis
Stenotyla helleri
*Stenotyla lankesteriana**
*Stenotyla lendyana** (type species of *Stenotyla*)
*Stenotyla picta**

* Molecular sampling confirms placement of this species within the genus
  Stenotyla.

Key to the Species of *Stenotyla*

1a. Inflorescence short, under 5 cm long, flower white, lip blotched with
purple, callus with four teeth . . . . *S. lankesteriana*

1b. Inflorescence over 5 cm long, flower color otherwise . . . . go to 2

2a. Flower concolor (all one color or shades of same color—yellow), callus
with two or more teeth . . . . *S. lendyana*

2b. Flower not concolor, lip with two distinct colors . . . . go to 3

3a. Callus with four apical teeth of equal length, flower yellow, with violet
longitudinal stripes interspersed with minute red-brown shading and
spots, column striped dark red-violet . . . . *S. helleri*

3b. Callus two-toothed or with two medial teeth and two shorter lateral
teeth . . . . go to 4

4a. Callus two-toothed. . . . . *S. picta*

4b. Callus two-toothed with two shorter lateral teeth . . . . *S. estrellensis*

Stenotyla estrellensis (Ames) P. A. Harding,
comb. nov. Basionym: *Chondrorhyncha
estrellensis* Ames, *Sched. Orchid.* 4: 54. 1923.
Type: Costa Rica, Cartago, Estrella, 10
January 1923, *Lankester & Sancho 396*
(holotype: AMES).

DESCRIPTION The flowers are white to yellow,
the lip has red longitudinal lines, and the callus is yellow. The dorsal sepal and petals are
erect. The lip is three-lobed, with undulate
margins, and the lip is strongly concave at its
base. The callus is deeply bilobed at the apex
with two small ancillary blunt teeth on each
side.

MEASUREMENTS Leaves 19 cm long, 1.5–2.5 cm
wide; inflorescence 13 cm long, with few bracts;
sepals 2.7 cm long, 0.5 cm wide; petals 2.5 cm
long, 0.9–1.0 cm wide; lip 2.5–3.0 cm long, 2 cm
wide; column 1.2 cm long.

COMMENT For a comparison of *Stenotyla estrellensis* and *S. helleri*, see discussion under *S.
helleri*.

ETYMOLOGY Named for the location where it
was originally found.

DISTRIBUTION AND HABITAT Known only from
Costa Rica (Nicaragua if you include *S. helleri*),
at 1500 m in elevation.

FLOWERING TIME January.

Stenotyla helleri (Fowlie) P. A. Harding, *comb. nov.* Basionym: *Chondrorhyncha helleri* Fowlie, *Orch. Digest* 35: 170. 1971. Type: Nicaragua, Matagalpa, 1500 m, *Heller 10097* (holotype: UCLA).

DESCRIPTION The flowers are cream-yellow, the lip is yellow with deep red-violet longitudinal lines interspersed with minute red-brown shading and spots, and the column is striped dark red-violet. The dorsal sepal and petals arch over the column forming a loose hood over it. The petals reflex at their apex. The lateral sepals reflex back. The lip is slightly three-lobed and tubular. The lateral lip lobes clasp the column, and the apical end of the lip is undulate. The callus is 0.7 cm wide with four teeth apically. PLATE 137.

COMMENT Franco Pupulin (2001) stated that *Chondrorhyncha helleri* is a synonym of *C. estrellensis*. The Kew World Checklist of Monocotyledons lists *Stenotyla helleri* as a synonym of *S. estrellensis*, which it does resemble, but the color of the callus is red-brown in *S. helleri*, yellow in *S. estrellensis*. In addition, the callus of *S. estrellensis* is more pointed apically, with definite side points, whereas the callus of *S. helleri* has four apical teeth that are described by Fowlie (1971) as being "shallowly excavated between the teeth." Both of these species could be also argued as being variations of *S. picta*, differing in color and other minor variations. I have decided to leave them as separate species as there are differences, minor though they may be.

MEASUREMENTS Leaves 25 cm long, 3.5 cm wide; inflorescence 15 cm long, with two bracts; dorsal sepal 2 cm long, 0.7 cm wide; lateral sepals 2.6 cm long, 0.7 cm wide; petals 2.2 cm long, 1 cm wide; lip 2.2 cm long, 1.9 cm wide; column 1 cm long.

ETYMOLOGY Named for A. H. Heller, who contributed much to the knowledge of Nicaraguan orchids.

DISTRIBUTION AND HABITAT Known only from Nicaragua, at 1500 m in elevation, in wet montane cloud forest.

FLOWERING TIME April.

Stenotyla lankesteriana (Pupulin) Dressler, *Lankesteriana* 5(2): 96. 2005. Basionym: *Chondrorhyncha lankesteriana* Pupulin, *Lindleyana* 15(1): 21. 2000. Type: Costa Rica, locality unknown, a confiscated plant flowered in cultivation at Lankester Botanical Garden, 19 May 1999, *F. Pupulin 1467* (holotype: USJ).

DESCRIPTION The flowers are white, with the lip blotched dark purple at the base; the blotches are slightly visible on the outer side of the lip. The elliptic lip is truncate with a short point at the apex. The apical lip margins are slightly undulate and minimally crisped. The callus is four-toothed at its apex. PLATE 138.

COMMENT There is an excellent photo by Franco Pupulin at www.neotrop.org/gallery labeled *Chondrorhyncha lankesteriana* and another at www.jardinbotanicolankester.org.

MEASUREMENTS Leaves 7–13 cm long, 1.0–1.2 cm wide; inflorescence 3.2–4.2 cm long, with two to four bracts; dorsal sepal 2.4 cm long, 0.8 cm wide; lateral sepals 2.6 cm long, 0.8 cm wide; petals 2.2 cm long, 0.9 cm wide; lip 2.7 long, 1.8 cm wide; column 1.2 cm long.

ETYMOLOGY Named for the botanical garden and institution where it was grown.

DISTRIBUTION AND HABITAT Known only from Costa Rica, based on one cultivated plant, elevation unknown.

FLOWERING TIME May.

Stenotyla lendyana (Reichenbach f.) Dressler, *Lankesteriana* 5(2): 96. 2005. Basionym: *Chondrorhyncha lendyana* Reichenbach f., *Gard. Chron.* 26: 103. 1886, as "*Chondrorrhycha*." (holotype: W); type species of *Stenotyla*.

Figure 5. *Stenotyla lendyana*. Drawing by Jane Herbst

SYNONYMS

Zygopetalum bidentatum Reichenbach f. ex Hemsley, *Biol. Cent. Amer.* 3(16): 251. 1884. Type: protologue Mexico, South Mexico, Chiapas, *Ghiesbreght 772* (holotype: K). *nom. nud.*

Warczewiczella bidentata (Reichenbach f. ex Hemsley) Schlechter, *Beih. Bot. Centralbl.* 36(2): 494. 1918.

Cochleanthes bidentata (Reichenbach f. ex Hemsley) R. E. Schultes & Garay, *Bot. Mus. Leafl.* 18: 323. 1959.

DESCRIPTION The flowers are yellow or white-yellow with a yellow tinge. The dorsal sepal is concave and carinate, and the lateral sepals are concave and reflex back. The petals have an undulate apical margin and deflex upward at their apical end. The lip is tubular, with the lateral margins clasping the column; the lip apex is undulate and rounded with a notch. The callus is flat, triangular, and bidentate or several-toothed. PLATE 139; FIGURE 5.

MEASUREMENTS Leaves 8–30 cm long, 1.2–4.0 cm wide; inflorescence 5.5–15.0 cm long, with two bracts; sepals 2–3 cm long, 0.5–0.7 cm wide; petals 2.0–2.5 cm long, 1.0–1.5 cm wide; lip 2.0–2.7 cm long and wide; column 1.0–1.3 cm long.

ETYMOLOGY Named for Major A. F. Lendy, who flowered the type specimen.

DISTRIBUTION AND HABITAT Known from Mexico to Panama, up to 1500–1700 m in elevation, in humid forests on tree trunks.

FLOWERING TIME June to August.

Stenotyla picta (Reichenbach f.) Dressler, *Lankesteriana* 5(2): 96. 2005. Basionym: *Warczewiczella picta* Reichenbach f., *Gard. Chron.* 20: 8. 1883. Type: Costa Rica, Endrés, *Pfau* (holotype: W).

SYNONYMS

Zygopetalum pictum (Reichenbach f.) Reichenbach f., *Gard. Chron.*, n.s., 20: 8. 1883, *in nota.*

Cochleanthes picta (Reichenbach f.) Garay, *Bot. Mus. Leafl.* 21: 256. 1967.

Warczewiczella caloglossa Schlechter, *Repert. Spec. Nov. Regni Veg.* 7: 216. 1913.

Chondrorhyncha caloglossa (Schlechter) P. H. Allen, *Ann. Missouri Bot. Gard.* 36: 85. 1949.

Chondrorhyncha picta (Reichenbach f.) Senghas, *Orchidee (Hamburg)* 41(3): 94. 1990.

DESCRIPTION The flowers are green-white, the lip is densely covered with red-purple lines on the apical half, and the callus is green. The dorsal sepal and petals hood over the lip and column, and the lateral sepals reflex somewhat. The lip is three-lobed and minimally crisped at its apical margin. The callus is small and bilobed or toothed at the callus apex. PLATE 140.

COMMENT There are excellent photos by Franco Pupulin and Diego Bogarín at www.neotrop.org/gallery labeled "*Chondrorhyncha picta*" and another by Franco Pupulin at www.jardinbotanicolankester.org.

MEASUREMENTS Leaves 30 cm long, 2.5 cm wide; inflorescence 8.5–10.0 cm long with one bract; dorsal sepal 2.5 cm long, 0.9 cm wide; lateral sepals 2.8 cm long, 0.8 cm wide; petals 2.7 cm long, 0.8 cm wide; lip 3 cm long, 2.8 cm wide; column 1.4 cm long.

ETYMOLOGY Latin *pictus*, pointed or colored, referring to the lip.

DISTRIBUTION AND HABITAT Known from Costa Rica and Panama, at 1400–2100 m in elevation, found in the shade on tree trunks in cloud forests.

FLOWERING TIME December to March.

Warczewiczella Reichenbach f., *Bot. Zeit. (Berlin)* 10: 635. 1852.

Warczewiczella, the last genus in this volume only because of alphabetical order, is probably one of the most common in cultivation, mostly because the plants are less particular about their growing conditions than the plants of other genera are. John Lindley in 1849 described *Warrea discolor*. Heinrich Gustav Reichenbach felt this species was better served by forming a new genus and created *Warczewiczella* in 1852, transferring *Warrea candida*, *W. discolor*, and *W. marginata* at the same time.

The new genus was named for Josef Warszewicz, who extensively collected and drew plants and was inspector of the Botanic Garden of Krakow. Reichenbach described the callus of *Warczewiczella* as being adnate to the base of the labellum but free laterally; and in comparing it to *Cochleanthes*, listed the traits of *Warczewiczella* as having a longer chin to the lip, greater reflection of its lateral sepals, a smaller more basal callus, and equal pollen masses.

Since that time *Warczewiczella* has gone in and out of favor as a useful genus. The most notable event in its history is Leslie Garay's article in 1969, in which he decided that all the species involved shared so many attributes that they would be better considered one genus *Cochleanthes*. Molecular data support the genus, with the exception of *W. wailesiana*, which lies outside of genus boundaries on the cladogram, but Robert Dressler (Whitten et al. 2005) combined *W. wailesiana* in *Warczewiczella* rather than form a new genus with only one species. The molecular data show *Warczewiczella* and *Cochleanthes* are clearly separate and definitely two distinct genera. The species of *Warczewiczella* are closer molecularly to *Aetheorhyncha*, *Chaubardiella*, *Chondroscaphe*, *Ixyophora*, and *Pescatorea*. The species of *Cochleanthes* are closest to *Stenotyla*.

Plants of *Warczewiczella* are generally moderately sized. The growth is fan-shaped, with the leaves being conduplicate, attached to the folded leaf petiole. The large flowers are presented on semierect to laxly pendent inflorescences, the flower/inflorescence height reaching the mid portion of the plant. The plants generally bloom during months with longer days, either blooming at one flush of flowers or several times per season. The flowers are a white-purple combination, with the lips being the most prominent feature. The lateral sepals reflex back (divergent) in most species. The lip is tubular only around the column with the lateral edges of the lip folding over or at least attempting to clasp the column. The medial lip lobe is large, functioning supposedly as an attractant or landing pad, being laxly tubular or flat for most of the lip length, with only the most apical part of the lip being deflexed. The lips have a variable length thick claw at their base, but often the length of the claw is so small as to not be present. The callus is basal, radiating or fanning both apically and laterally with (in most of the species) thick prominent keels that create a raised plate. The column is widest at mid length, has no to minimal wings, and has a small tooth on the short ventral column foot. The column foot is held at an approximate right angle to the column, but this is variable. The stigma is slitlike, the rostellum is toothlike. The pollinia are equal, attached to a short shield-shaped stipe and a shield-shaped viscidium.

Some of the above traits are shared with the species of *Chondrorhyncha* and *Kefersteinia*; however, the flowers of *Kefersteinia* do not have reflexed lateral sepals. *Chondrorhyncha* and *Kefersteinia* have a very indistinct claw on the lip compared to the distinctly thickened claw of *Cochleanthes* and *Warczewiczella*.

Etymology

Named by Heinrich Gustav Reichenbach for Josef Warszewicz (var-SHEV-itch), who extensively collected and drew plants, but was also inspector of the Botanic Garden of Krakow. Reichenbach misspelled the name as *warscewiczii* for most of the species he named after Warszewicz, but spelled it a little more correctly when he named the genus *Warczewiczella*. The Kew World Checklist of Monocotyledons lists the spelling as *Warczewiczella*, w³TROPICOS has the genus listed with two spellings, and the International Plant Name Index has it as *Warczewiczella* though a few species are spelled *Warscewiczella*.

List of the Species of *Warczewiczella*

Warczewiczella amazonica

Warczewiczella candida

*Warczewiczella discolor** (type species of *Warczewiczella*)

*Warczewiczella guianensis**

Warczewiczella ionoleuca

*Warczewiczella lipscombiae**

Warczewiczella lobata

*Warczewiczella marginata**

Warczewiczella palatina

Warczewiczella timbiensis

*Warczewiczella wailesiana***

* Molecular sampling confirms placement of this species within the genus *Warczewiczella*.

** This species kept in the genus *Warczewiczella* despite molecular sampling results. See text for further details.

Key to the Species of *Warczewiczella*

1a. Lip one-lobed, or lateral lobes obscure and blunt, the part that folds toward the column created by folding of lateral edges of the lip, not exactly lateral lobes of lip. . . . . go to 2

1b. Lip distinctly three-lobed. . . . . go to 6

2a. Flowers large, more than 6 cm across, white, lip white with purple veins. . . . .*W. amazonica*

2b. Flowers smaller than 6 cm across, flower color otherwise. . . . . go to 3

3a. Flower open and flat, lateral margins or lobes of lip do not cover the column. . . . . go to 4

3b. Flower tubular or bell-shaped, lateral margins of lip cover or touch the column. . . . . go to 5

4a. Flower white, margin of lip pink or purple . . . . *W. candida*

4b. Flower cream-white, lip callus and veins lavender centrally . . . *W. wailesiana*

5a. Lip purple, not just at margins or veins . . . .*W. discolor*

5b. Lip white with pink margin and pink-red veins centrally. . . . . *W. guianensis*

6a. Lateral lip lobes large, at least half the size of the lip midlobe, lip tubular . . . . go to 7

6b. Lateral lip lobes small, acute, lip blade held mostly flat . . . . go to 9

7a. Callus platelike without obvious keels or teeth, flower white sometimes with red lip margin . . . .*W. lobata*

7b. Callus with obvious keels and teeth . . . . go to 8

8a. Lip with pink or red margins, no purple veins . . . . *W. ionoleuca*

8b. Lip inner surface with purple veins . . . . *W. palatina*

9a. Flower white with purple veins, no other coloration on flower . . . .*W. amazonica*

9b. Flower color otherwise . . . . go to 10

10a. Flower green, lip white with deep purple veins to the margins of the lip. . . . .*W. timbiensis*

10b. Flower white, veins and margin of lip pink or purple . . . . go to 11

11a. Lip with central purple spot apical to the callus. . . . . *W. marginata*

11b. Lip with red or purple-red veins, no spot of color apical to the callus . . . . *W. lipscombiae*

Warczewiczella amazonica Reichenbach f. & Warszewicz, *Bonplandia* 2: 97. 1854. Type: Peru, upper Amazon regions, 1853, *Warszewicz* (holotype: W).

SYNONYMS

Zygopetalum amazonicum (Reichenbach f. & Warszewicz) Reichenbach f., *Ann. Bot. Syst.* 6: 655. 1863.

Warczewiczella lindenii (Rolfe) auct., *J. Hort. Pract. Gard.* 1: 419. 1892, as *Warscewiczella*.

Zygopetalum lindenii Rolfe, *Lindenia* 8: 5, plate 337. 1892.

Cochleanthes amazonica (Reichenbach f. & Warszewicz) R. E. Schultes & Garay, *Bot. Mus. Leafl.* 18: 322. 1959.

Chondrorhyncha amazonica (Reichenbach f. &

Warszewicz) Hawkes, *Enc. Cult. Orchids* 100. 1965.

DESCRIPTION The flowers are white with purple veins in the lip. The dorsal sepal and petals curl back only at the apex. The lateral sepals strongly reflex back. The lip is one- to three-lobed though the lateral lobes may be obscure. The lateral lobes clasp but do not cover the column and the midlobe is large with undulate margins. The callus plate is rhomboid, not extending to the lateral lip lobes, with two distinct teeth or points at the apex of the callus. The column is mostly straight, the clinandrium is covered apically by the column apex, the stigma is a transverse slit, and the column foot is small and toothed. The pollinia are equal on a small stipe, and the viscidium is shield-shaped. PLATE 141.

COMMENT *Warczewiczella amazonica* is said to be very similar to *W. lipscombiae* (from Panama) but differs in having larger flowers. See discussion under *W. lipscombiae* for additional details. *Warczewiczella amazonica* has been line bred and the measurements following are for wild-collected average-sized plants.

MEASUREMENTS Leaves 19–25 cm long, 2.0–2.5 cm wide; inflorescence 4 cm long; sepals and petals 3 cm long, 1 cm wide; lip 4 cm long, 3 cm wide; column 2 cm long.

ETYMOLOGY Named for the region in which the plant was first found.

DISTRIBUTION AND HABITAT Known from Brazil, Colombia, Ecuador, Peru, and Suriname, at 500–1000 m in elevation, in wet forests.

FLOWERING TIME April to November.

Warczewiczella candida Reichenbach f., *Bot. Zeit. (Berlin)* 10: 636. 1852. Basionym: *Warrea candida* Lindley & Paxton, *Flow. Gard.* 1: 32. 1850. Type: Brazil, Bahia (holotype: K).

SYNONYMS
Huntleya candida (Lindley) Walpers, *Ann. Bot. Syst.* 3: 543.1852.

Cochleanthes candida (Lindley) R. E. Schultes & Garay, *Bot. Mus. Leafl.* 18: 323. 1959.
Zygopetalum candidum (Lindley) Reichenbach f., *Ann. Bot. Syst.* 6: 656. 1863.

DESCRIPTION The flower is white, the lip is white with a blue-violet center, the apical margin of the lip is white suffused with red to blue-lavender. The lip is ovate, notched at the apex, sometimes with a small central point at the apex. The callus has five lamellae that spread and are raised off the base of the lip apically. PLATE 142.

COMMENT The flower is very similar to that of *Warczewiczella wailesiana*. In fact, the easy way to describe the difference between the two is that *W. candida* is less white, having a pigmented lip margin. *Orchid Digest* (1969, 33) pictures the two side by side: *W. candida* spreads its five-lamellae callus more like a star, and its lateral lip edges don't go above the column, whereas the *W. wailesiana* five-lamellae callus is more adpressed to the lip and the lateral lip edges go above the column.

MEASUREMENTS Dorsal sepal 2.5 cm long, 1 cm wide; lateral sepals 2.3 cm long, 0.9 cm wide; petals 2.2 cm long, 0.8 cm wide; lip 2.2 cm long, 2.3 cm wide.

ETYMOLOGY Latin *candidus*, pure glossy white, describing the color of the tepals.

DISTRIBUTION AND HABITAT Known from western Brazil in Minas Gerais, Bahia, Espírito Santo, and Rio de Janeiro.

FLOWERING TIME Not recorded.

Warczewiczella discolor (Lindley) Reichenbach f., *Bot. Zeit. (Berlin)* 10: 636. 1852. Basionym: *Warrea discolor* Lindley, *J. Hort. Soc.* 4: 265. 1849. Type: Costa Rica, Warszewicz (holotype: K); type species for *Warczewiczella*.

SYNONYMS
Zygopetalum discolor Reichenbach f., *Ann. Bot. Syst.* 6: 655. 1861.

Chondrorhyncha discolor (Lindley) P. H. Allen, *Ann. Missouri Bot. Gard.* 36: 87. 1949.

Cochleanthes discolor (Lindley) R. E. Schultes & Garay, *Bot. Mus. Leafl.* 18: 324. 1959.

DESCRIPTION The sepals are white to yellow-white, sometimes green at the apex, the petals are white, sometimes suffused with purple toward the apex, the lip is deep violet-purple, sometimes with white margins, and the callus is lavender, white, or dark yellow. The inflorescence is suberect. The obscurely to conspicuously three-lobed lip has lateral lobes that are erect and incurve over the column. The callus radiates like a fan with fingerlike projections, the free projecting apex of the callus has elongate teeth of unequal length. The column is stout, hirsute on the ventral surface, with a ventral tooth on the column foot. The viscidium is shield-shaped.

COMMENT This species is widely seen in cultivation, and many different forms are in cultivation. The Internet has many, many photos of this species, and it is seldom misidentified.

MEASUREMENTS Leaves 20–25 cm long, 2.5–4.5 cm wide; dorsal sepal 2.5 cm long, 1.2 cm wide; lateral sepals 2.7 cm long, 1.1–1.2 cm wide; petals 2.5 cm long, 1.4 cm wide; lip 3 cm long, 2.8–3.2 cm wide; column 1.5 cm long.

ETYMOLOGY Latin *discolor*, not of the same color, referring to the colors of the sepals and petals and lip.

DISTRIBUTION AND HABITAT Known from Cuba, Honduras, Costa Rica, and Panama through Colombia and Venezuela and south to Ecuador and Peru. A common inhabitant of the cooler creek forests in the Upper Pacific slopes of Costa Rica and Panama at 1900–2300 m in elevation. In Venezuela it is known from 1700 m in elevation. The species grows on moss-covered trunks and lower branches, particularly in creek gulches, where plants receive high humidity by night condensation.

FLOWERING TIME Early to late spring.

Warczewiczella guianensis (A. Lafontaine, G. Gerlach & Senghas) Dressler, *Lankesteriana* 5(2): 96. 2005. Basionym: *Cochleanthes guianensis* A. Lafontaine, G. Gerlach & Senghas, *Orchidee (Hamburg)* 42(6): 285. 1991. Type: French Guiana, 200 m, collected by and flowered in cultivation by A. Lafontaine, *Botanical Garden Heidelberg O-20030* (holotype: HEID).

DESCRIPTION The sepals and petals are white, the lip is white with a pink margin and pink-red veins, and the callus has pink veins. The dorsal sepal and petals curl back at their apexes, the lateral sepals are reflexed. The lip is three-lobed, the lateral lobes are small and obscure if not lacking. The lateral edges of the lip touch but do not cover the column. The lip midlobe has a broad notch at the apex. The callus has 14 ridges and is slightly toothed apically. The column is widest at the stigma and is hirsute on the ventral surface. PLATE 143.

COMMENT *Warczewiczella guianensis* looks a lot like *W. marginata*, but only the lateral edges of the lip of *W. guianensis* fold inward whereas *W. marginata* has definite lateral lobes.

MEASUREMENTS Leaves 20 cm long, 5 cm wide; inflorescence 10 cm long; dorsal sepal 3.3 cm long, 1.6 cm wide; lateral sepals 3.8 cm long, 1.6 cm wide; petals 3.5 cm long, 1.8 cm wide; lip 4.5 cm long, 4 cm wide; column 1.5 cm long.

ETYMOLOGY Named for the country in which the original specimen was found.

DISTRIBUTION AND HABITAT Known only from French Guiana, at 200 m in elevation.

FLOWERING TIME Not recorded.

Warczewiczella ionoleuca Schlechter, *Repert. Spec. Nov. Regni Veg. Beih.* 7: 267. 1920. Basionym: *Zygopetalum ionoleucum* Reichenbach f., *Bot. Zeit. (Berlin)* 23: 99. 1865. Type: Colombia (holotype: W).

SYNONYM

Cochleanthes ionoleuca (Reichenbach f.) R. E.

Schultes & Garay, *Bot. Mus. Leafl.* 18: 325. 1959.

DESCRIPTION The sepals and petals are green-yellow, the lip is pink or yellow with a pink margin, the callus is a darker pink or yellow with brown veins, and the column is pink. The sepals and petals are reflexed at their apexes. The lip is three-lobed, the lateral lobes are narrow, erect, touching but not covering the column. The callus of about 15 keels or ridges is raised, the apical margin of the callus has teeth or points. The column has two minimal wings on each side, one small lobe lateral to the stigma, the other lobe at mid length of the column. PLATE 144.

COMMENT *Warczewiczella ionoleuca* differs from *W. lobata* in having a callus with ridges or keels whereas the callus of *W. lobata* is a flat disc. The shape of the lip is similar for the two species, but is narrower in *W. ionoleuca* and broader in *W. lobata*. There is a lovely photo of *W. ionoleuca* by Franco Pupulin at www.jardinbotanicolankester.org.

MEASUREMENTS Leaves 18 cm long, 3 cm wide; inflorescence 9 cm long; dorsal sepal 3.5 cm long, 1 cm wide; lateral sepals 4 cm long, 1 cm wide; petals 3.5 cm long, 1 cm wide; lip 3 cm long, 3.5 cm wide; column 0.8 cm long.

ETYMOLOGY Greek *ion*, violet-colored, and *leuca*, white, referring to the color of the lip.

DISTRIBUTION AND HABITAT Known only from Colombia and Ecuador, at 850 m in elevation.

FLOWERING TIME Not recorded.

Warczewiczella lipscombiae (Rolfe)

Fowlie, *Orch. Digest* 33: 229. 1969. Basionym: *Chondrorhyncha lipscombiae* Rolfe, *Bull. Misc. Inform. Kew*: 133. 1912. Type: Panama (holotype: K).

SYNONYM
Cochleanthes lipscombiae (Rolfe) Garay, *Orquideología* 4: 152. 1969.

DESCRIPTION The flowers are white with purple veins in the lip and a pink suffusion on the lip margin. The lip is three-lobed, the lateral lobes of the lip are erect and incurve over the column, the central lip lobe is notched at the apex. The callus is a square disc with many short apically projecting teeth. All the teeth are about equal in length. PLATE 145.

COMMENT *Warczewiczella lipscombiae* is very similar in appearance to *W. marginata* but lacks the blush of color apical to the callus in the middle of the lip and has very light to minimal suffusion on the lip margin compared to some plants of *W. marginata* where the color is deep. *Warczewiczella lipscombiae* is also said to be very similar to *W. amazonica* (from South America) but differs in having smaller flowers, lateral lip lobes, and infolded lateral lip edges, whereas *W. amazonica* has larger flowers, lacks lateral lip lobes, and forms its upcurled edges by infolding the lateral edges of the lip.

MEASUREMENTS None recorded.

ETYMOLOGY Named for Mrs. Lipscomb, who sent Robert Rolfe the plant collected by her son Lancelot near Las Cascades and also near Bohio in Panama.

DISTRIBUTION AND HABITAT Known only to Panama, found in damp shaded locations at low elevations (near sea level).

FLOWERING TIME Not recorded.

Warczewiczella lobata (Garay) Dressler,

Lankesteriana 5(2): 96. 2005. Basionym: *Cochleanthes lobata* Garay, *Orquideología* 4: 21. 1969. Type: Colombia, cultivated by Alvaro Mejía s.n. (holotype: AMES).

DESCRIPTION The flowers are green-white. The dorsal sepal is erect and slightly concave. The lateral sepals are reflexed. The petals covering the column are reflexed slightly at their apexes. The lip is three-lobed, the lateral lobes are erect and encircle the column, and the lip midlobe is kidney-shaped and further divided into four smaller lobes and notched at the apex. The callus is a raised flat disc without apical teeth. PLATE 146.

COMMENT The general appearance of the flower of *Warczewiczella lobata* is similar to that of *W. ionoleuca*, but the drawing of the type of *W. lobata* shows the apexes of the lateral lip lobes to be more acute than those of *W. ionoleuca*. Furthermore, *W. lobata* is smaller, lacks keels in the callus, and has a different flower color pattern.

MEASUREMENTS Leaves 22 cm long, 2.5 cm wide; inflorescence 10 cm long; dorsal sepal 2.5 cm long, 1.2 cm wide; lateral sepals 3 cm long, 1 cm wide; petals 2.5 cm long, 1 cm wide; lip 2.5 cm long, 3.5 cm wide; column 1.5 cm long.

ETYMOLOGY Latin *lobatus*, lobed, referring to the multilobed lip.

DISTRIBUTION AND HABITAT Known only from Colombia, elevation not recorded.

FLOWERING TIME Not recorded.

Warczewiczella marginata Reichenbach f., *Bot. Zeit. (Berlin)* 10: 636. 1852. Type: W.

SYNONYMS
Warrea marginata Reichenbach f., *Bot. Zeit. (Berlin)* 10: 636. 1852, *pro syn.*
Huntleya marginata hort. ex Reichenbach f., *Bot. Zeit. (Berlin)* 10: 636. 1852, *pro syn.*
Warrea quadrata Lindley, *Gard. Chron.*: 647. 1853.
Warczewiczella velata Reichenbach f. & Warszewicz, *Bonplandia* 2: 97. 1854.
Zygopetalum marginatum Reichenbach f., *Ann. Bot. Syst.* 6: 654. 1861.
Zygopetalum velatum (Reichenbach f. & Warszewicz) Reichenbach f., *Ann. Bot. Syst.* 6: 655. 1863.
Zygopetalum quadratum Pfitzer, *Vergl. Morph. Orch.*: 58. 1881.
Chondrorhyncha marginata (Reichenbach f.) P. H. Allen, *Ann. Missouri Bot. Gard.* 36: 88. 1949.
Cochleanthes marginata (Reichenbach f.) R. E. Schultes & Garay, *Bot. Mus. Leafl.* 18: 326. 1959.

Zygopetalum fragrans Linden, *Cat. Gén.* 6: 9. 1851. *quo vide*. See Fowlie 1969b.

DESCRIPTION The sepals and petals are green-white, the lip is white with lavender to pink radiating lines and a central overflush of purple just apical to the callus, the central portion of the lip blade is white, and the lip margin has a broad confluent flush of pink-lavender. The flower color is highly variable but the flower pattern is not. The lateral sepals are conduplicate, spreading, and divergent. The lip is three-lobed, the lateral lobes of the lip are erect and incurve over the column, the medial lobe is notched at the apex. The callus is spreading, fleshy, and with many lamellae or keels with short apically projecting teeth. All the teeth are about the same length. The flowers are fragrant of cinnamon or cloves in the morning hours. PLATE 147.

COMMENT *Warczewiczella marginata* looks a lot like *W. guianensis*, but *W. marginata* has definite lateral lobes, whereas only the lateral edges of the lip of *W. guianensis* fold inward.

MEASUREMENTS Leaves 12–30 cm long, 2.0–4.5 cm wide; inflorescence 5.5–10.0 cm long; dorsal sepal 2.2–3.0 cm long, 0.9–1.6 cm wide; lateral sepals 2.9–3.2 cm long, 0.8–1.6 cm wide; petals 2.5–3.0 cm long, 1.0–1.5 cm wide; lip 3.5–4.5 cm long, 3.0–4.4 cm wide; column 1.2–1.8 cm long.

ETYMOLOGY Latin *marginatus*, margin, referring to the showy colored margin on the lip.

DISTRIBUTION AND HABITAT Known from Columbia, Panama, and Venezuela, at 1300 m in elevation.

FLOWERING TIME Not recorded.

Warczewiczella palatina (Senghas) Dressler, *Lankesteriana* 5(2): 96. 2005. Basionym: *Cochleanthes palatina* Senghas, *Orchidee (Hamburg)* 41(3): 96. 1990. Type: Bolivia, La Paz, 1000 m, collected by S. & K. Lefernz, *Botanischer Garten Heidelberg, sub Orch-691y* (holotype: HEID).

DESCRIPTION The sepals and petals are pale cream-yellow with barely discernible pale green-yellow veins, the lip exterior is white tinted pale yellow, the lip inner surface is cream-yellow, the veins of the lip are purple, the lip margins are dull purple, the callus is yellow with dull purple lines or veins, the column is white dorsally, the foot and claw are bright yellow, the anther cap is white, and the pollinia are pale yellow. The lip is three-lobed, the apical margin of the median lobe is revolute. The transverse, thick callus is broadest across the base, tapering abruptly to the three-toothed medial apex. The column is elliptic in overall shape from below, covered with a short pubescence, the apex of the column is arcuate with three thick lobes surrounding the anther, the lateral lobes separated by a deep furrow in which lies the strap-shaped rostellum, and the sides of the column have four or five obtuse swellings or short ridges. The anther cap is papillose. PLATE 148.

MEASUREMENTS Leaves 15–30 cm long, 1–2 cm wide; inflorescence 6–8 cm long; dorsal sepal 3 cm long, 1.4 cm wide; lateral sepals 3 cm long, 1.1 cm wide; petals 2.8 cm wide, 1.3 cm long; lip 2.5 cm long, 1.8 cm wide; column 2 cm long, 0.6 cm wide.

ETYMOLOGY Latin *palatum*, the roof of the mouth, referring to the appearance of the lip. A German organization known as Orchideen-Gesellschaft Kurpfalz e.V. (Orchid Society Palatinat) claims the orchid is named after it. Palatinate is a region in the Upper River Rhine Valley.

DISTRIBUTION AND HABITAT Known from Bolivia and Peru, at 1000 m in elevation, in wet montane forest.

FLOWERING TIME January to April.

Warczewiczella timbiensis P. Ortíz,
Orquideología 24(1): 7. 2005. Type: Colombia, Cauca, Timbío, 1700 m, *P. Ortíz 1229* (holotype: HPU).

DESCRIPTION The sepals and petals are yellow-green, the lip is white with prominent purple-blue veins that are more dense at the margins. The lip is three-lobed, the lateral lobes enfold the column, the central lobe is broad with an undulate margin. The callus is quadrate, the apex is slightly raised and bluntly three-pointed. The column is enlarged at the apex. PLATE 149.

COMMENT This species has been collected only once in an area where no other *Warczewiczella* or *Cochleanthes* species are present. The area has been walked over several times looking for more plants or relatives. Still this species could possibly be a natural hybrid, as hybrids between *Cochleanthes* and *Warczewiczella* often result in strikingly colored lips. Crossing this plant to itself would potentially solve this question.

MEASUREMENTS Leaves 24 cm long; inflorescence 5 cm long, with five bracts; dorsal sepal 3 cm long, 1 cm wide; lateral sepals 3 cm long, 0.7 cm wide; petals 3 cm long, 1.4 cm wide; lip 4 cm long, 4.5 cm wide; column 1 cm long.

ETYMOLOGY Named after the town where the plant was found.

DISTRIBUTION AND HABITAT Known only from Colombia, at 1700 m in elevation.

FLOWERING TIME Not recorded.

Warczewiczella wailesiana (Lindley)
E. Morren, *Ann. Hort. Belg.* 28: 183. 1878. Basionym: *Warrea wailesiana* Lindley, *J. Hort. Soc.* 4: 264. 1849. Type: Brazil (holotype: K).

SYNONYMS
Warrea digitata Lemaire, *Ill. Hort.* 3, misc. 70. 1856.
Zygopetalum wailesianum (Lindley) Reichenbach f., *Ann. Bot. Syst.* 6: 656. 1863.
Warczewiczella digitata (Lemaire) Barbosa Rodrigues, *Struct. des Orchid.*: 13. 1883.
Cochleanthes wailesiana (Lindley) R. E. Schultes & Garay, *Bot. Mus. Leafl.* 18: 326. 1959.

Cochleanthes digitata (Lemaire) R. E. Schultes & Garay, *Bot. Mus. Leafl.* 18: 324. 1959.
Chondrorhyncha wailesiana (Lindley) A. D. Hawkes, *Enc. Cult. Orchids*: 111. 1965.

DESCRIPTION The sepals and petals are green, white, or cream, and the lip is white with violet stains along the veins and callus. The callus has five spreading fingerlike bars, which are free except at the base. The flower has the fragrance of sweet peas or hyacinths. PLATE 150.

COMMENT This flower of *Warczewiczella wailesiana* is much like that of *W. candida* but has more green in the sepals and petals and lacks the pink lip margin of *W. candida*. There is very little difference in shape, and since both species are variable in their color forms, some consider *W. candida* and *W. wailesiana* synonyms, in which case the name *W. wailesiana* would have priority.

MEASUREMENTS Leaves 15–30 cm long, 2–3 cm wide; inflorescence 8–12 cm long; sepals 2.5 cm long, 0.9–1.0 cm wide; petals 2.2 cm wide, 1.0–1.2 cm wide; lip 2.7–3.0 cm long, 2.8–3.6 cm wide; column 1.1 cm long, 0.5 cm wide.

ETYMOLOGY Named for G. Wailes, who flowered the type specimen.

DISTRIBUTION AND HABITAT Known only from Brazil, at 500 m in elevation.

FLOWERING TIME Not recorded.

Unplaced Species Names

Several species names have come up, yet I am not sure what do with them.

Chondrorhyncha macronyx Kraenzlin, *Notizbl. Bot. Gart. Berl.* 7: 414. 1920, as *"Chondrorrhyncha."* Type: Colombia, Antioquia, at 17 m, *Kalbreyer 1755* (holotype: not located, perhaps Berlin and destroyed in 1944).

DESCRIPTION The flowers are white with brown spots. The lip is spathe-shaped, basally cordate with a long (0.9 cm) claw, mildly curved, wavy and crenulate or fimbriate. The callus is two-lobed, each side with two parts, and is minutely pubescent. The column is stout and clavate.

COMMENT This has previously been classified as *Chondrorhyncha*, and Whitten et al. (2005) list it as a *Chondrorhyncha sensu stricto*. None of the other species included in *Chondrorhyncha sensu stricto* have fimbriate or crenulate lips or long claws to the lip, or particularly two-lobed calluses with two parts each. Based on the length of the claw from the column foot to the fan-shaped blade of the lip and the two-lobed callus, I feel this species more likely fits into *Kefersteinia*, though I have not seen the flower or a drawing. Three species could fit this description, and perhaps the closest fit would be *Kefersteinia tolimensis*, if the specimen was one of the inferior forms with narrow lips and a longer than normal claw, or perhaps *K. taggesellii*. Not finding the specimen, or having anything else but the written description to go on, I feel it is best left in this chapter.

MEASUREMENTS Leaves 25–30 cm long, 1.8 cm wide; inflorescence 7 cm long; sepals and petals 2.2–2.5 cm long, 0.6–0.7 cm wide; lip blade 1 cm long, 1.5 cm wide.

ETYMOLOGY Named after *Macronyx*, a genus of birds with unusually long claws, referring to the long spur on the rostellum of this plant.

DISTRIBUTION AND HABITAT Known from Colombia, at low elevations.

FLOWERING TIME June.

Kefersteinia leucantha Reichenbach f. ex Linden, *Ill. Hort.* 29: 52. 1882, *nom. nud.*

Venezuela. This name is invalid, but I have been unable to find out why or what species is represented.

Glossary

acuminate having a long tapering point; starting to narrow at mid length and tapering to a point

acute having a distinct or sharp point, not drawn out

alba white

albanistic yellow, with some pigments usually yellow

anchoriform anchor-shaped

apiculate ending in a short sharp point; like acuminate but with a blunter point

apicule a point

appressed lying flat on the a surface

arcuate bent like a bow; arching

asperity a slight projection from a surface; a point or bump

attenuate narrowed

basilar located at or near the base

bialate having two winglike parts

bidentate two-toothed

bifurcate to divide or separate into two parts

bilobed two-lobed

biseriate in two rows or portions

caespitose tufted

carinate keeled

caudate having a tail

clavate club-shaped; widest apically

cochleate shaped like a cockleshell

colliculate having rounded swellings

conduplicate folded lengthwise

cordiform heart-shaped

crenulate wavy; finely edged with rounded teeth that point forward

crispate irregularly toothed, waved, and twisted; looking like crumpled paper

cuneate wedge-shaped

decurrent having the leaf base extending down the stem as two wings

deflexed bent downward at a sharp angle

dentate toothed

denticulate finely toothed

distichous arranged alternately in two vertical rows on opposite sides of the stem

emarginate notched at the apex

entire without notches or cuts

epichil the apical portion of the lip

erose scalloped; irregularly chewed or bitten; unevenly gnawed

excavate carved out

falcate sickle-shaped

filiform threadlike

fimbria (pl. fimbriae) a fringelike part

fimbriate fringed

flabellate fan-shaped

foliaceous leaflike

geniculate elbowed; bent like an elbow

gibbose swollen, pouched or convex above, flattened below

glabrous without hairs

granulate grainy, granular

hemicircular half circular

hirsute covered with coarse, dense hairs

hypochil the basal portion of the lip

imbricating braided, overlapping

immaculate having no spots

involute having upward- or inward-rolled margins

lacerate having irregularly torn edges

lacinia (pl. **laciniae**) an irregular part of a petal

laciniate jagged, fringed

lamella (pl. **lamellae**) a small thin platelike structure

laminar platelike

ligulate strap-shaped

lobulate having lobules

lobules small earlike lobes

lunate crescent-shaped

mucronulate having an obtuse apex that ends in an abrupt, small, short, sharp tip

multilobulate having multiple small earlike lobules

multiplicate folded into pleats

oblanceolate reverse lanceolate, two or three times as long as broad, widest in the middle or apical portion of leaf and tapering to a pointed apex

obovate egg-shaped; widest at the base and gradually coming to a point

obpyriform pear-shaped, widest apically

obtuse rounded or blunt

operculate possessing a lid

ovate oval in shape but slightly widest below the middle

pandurate fiddle-shaped

papillose warty; having short protuberances

pilose softly hairy

plicate folded like a fan

porrect extending forward

puberulent downy pubescent

pulverulent looking as if covered by dust

pustule a small blister or swelling

pyriform pear-shaped

quadrate square

recurvate curving back on itself

reniform kidney-shaped

resupinate having the lip uppermost

retrorse bent backwards

retuse having a bluntly rounded end with a central notch

revolute curled under or out

rhombiform shaped like a parallelogram

rugulose warty

saccate pouched

sacciform pouched

sagittate shaped like an arrowhead with two barbs pointing backward

semicircular C-shaped

semiterete half cylindric

serrulate having small pointed teeth projecting forward, like a saw blade

sessile lacking a stalk or claw

sigmoid S-shaped

sinuate divided into wide irregular teeth or lobes that are separated by shallow notches

spatulate spoon-shaped

sphalm a Latin abbreviation meaning "slip of the pen"

subcrenulate finely scalloped; edged with rounded teeth that point forward

suberect not entirely erect

subflabellate mostly fan-shaped

suborbicular mostly orbicular

subquadrate almost square when flattened

subsaccate almost saccate, almost pouched

subsessile mostly sessile

sulcate having deep, narrow furrows

terete cylindrical

tessellated having a pattern following veins and cross striations

transverse across the surface long axis, not lengthwise

trichome a hairlike growth

trullate trowel-shaped

truncate blunt ended, as if cut off suddenly

tuberculate warty

undulate wavy

unguiculate narrowed into an elongate claw or straplike base

ventricose bulging on one side or in one direction

verrucose covered with warts

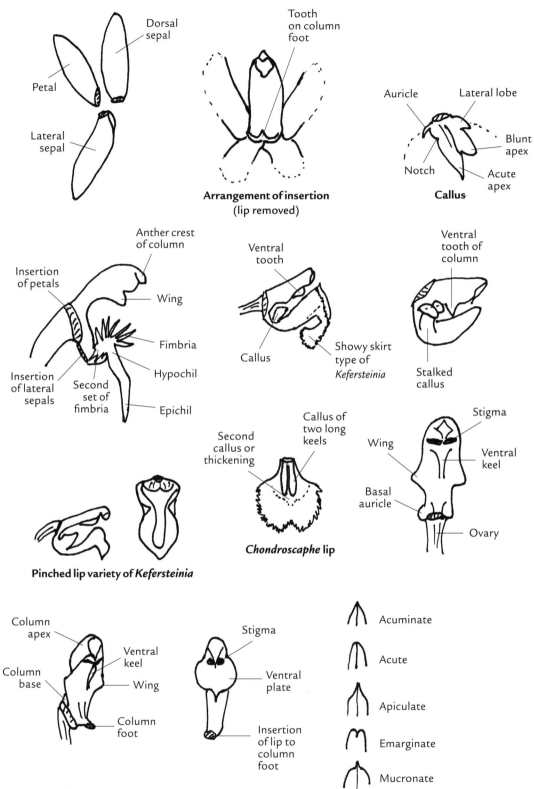

Arrangement of insertion
(lip removed)

Callus

Pinched lip variety of Kefersteinia

Chondroscaphe **lip**

Figure 6. Plant parts and shapes.

Petal
Dorsal sepal
Lateral sepal

Tooth on column foot

Auricle
Lateral lobe
Blunt apex
Acute apex
Notch

Anther crest of column
Insertion of petals
Wing
Fimbria
Hypochil
Epichil
Insertion of lateral sepals
Second set of fimbria

Ventral tooth
Callus
Showy skirt type of *Kefersteinia*

Ventral tooth of column
Stalked callus

Second callus or thickening
Callus of two long keels

Stigma
Wing
Ventral keel
Basal auricle
Ovary

Column apex
Ventral keel
Wing
Column base
Column foot

Stigma
Ventral plate
Insertion of lip to column foot

Acuminate
Acute
Apiculate
Emarginate
Mucronate

Selected References

Below is a list of publications used in the manuscript that the author suggests as further reading or references. It does not include all publications used for the research, as most used are mentioned in the text in reference to a specific species or group of plants.

Ackerman, James D. 1983. Euglossine bee pollination of the orchid *Cochleanthes lipscombiae*: a food source mimic. *American Journal of Botany* 70 (6): 830–834.

Ames, Oakes, and Donovan S. Correll. 1952. *Orchids of Guatemala*. Chicago: Chicago Natural History Museum.

Ames, Oakes, and Donovan S. Correll. 1985. *Orchids of Guatemala and Belize*. New York: Dover Publications.

Baker, Margaret, and Charles Baker. 1991. *Orchid Species Culture: Pescatorea, Phaius, Phalaenopsis, Pholidota, Phragmipedium, Pleione*. Portland, Oregon: Timber Press.

Bechtel, Helmut, Phillip Cribb, and Edmund Launert. 1986. *The Manual of Cultivated Orchid Species*. Cambridge, Massachusetts: MIT Press.

Béhar, Moises, and Otto Tinschert. 1998. *Guatemala y sus Orquídeas* (Guatemala and its orchids). Guatemala City: MayaPrint/Trade Litho.

Bennett, David E. 1998. *Icones Orchidacearum Peruviarum [3]*. Plates 401–600. Sarasota, Florida: A. Pastorelli de Bennett.

Bennett, David E. 2001. *Icones Orchidacearum Peruviarum [4]*. Plates 601–800. Sarasota, Florida: A. Pastorelli de Bennett.

Bennett, David E. 2008. *Icones Orchidacearum Peruviarum*. Sarasota, Florida: A. Pastorelli de Bennett.

Bennett, David E., Jr., and Eric A. Christenson. 1995. *Icones Orchidacearum Peruviarum*. Sarasota, Florida: A. Pastorelli de Bennett.

Brieger, F. G., R. Maatsch, and K. Senghas, eds. 1977. Schlechter's *Die Orchideen*. 3rd ed. Berlin and Hamburg: Verlag Paul Parey.

Dix, Margaret A., and Michael W. Dix. 2000. *Orchids of Guatemala: A Revised Annotated Checklist*. St. Louis, Missouri: Missouri Botanical Garden.

Dodson, Calaway H. 1993. *Native Ecuadorian Orchids*. Vol. 1, *Aa–Dracula*. Medellín, Colombia: Editorial Colina.

Dodson, Calaway H. 2001. *Native Ecuadorian Orchids*. Vol. 2, *Dresslerella–Lepanthes*. Medellín, Colombia: Editorial Colina.

Dodson, Calaway H. 2002. *Native Ecuadorian Orchids*. Vol. 3, *Lepanthopsis–Oliveriana*. Medellín, Colombia: Editorial Colina.

Dodson, Calaway H. 2003. *Native Ecuadorian Orchids*. Vol. 4, *Oncidium–Restrepiopsis*. Medellín, Colombia: Editorial Colina.

Dodson, Calaway H. 2004. *Native Ecuadorian Orchids*. Vol. 5, *Rodriquezia–Zygosepalum*. Medellín, Colombia: Editorial Colina.

Dodson, Calaway H., and David E. Bennett, Jr. 1989. *Orchids of Peru. Icones Plantarum Tropicarum*, series 2, fascicle 1. St. Louis, Missouri: Missouri Botanical Garden.

Dodson, Calaway H., and Piedad Marmol de Dodson. 1980. *Orchids of Ecuador. Icones Plantarum Tropicarum*, series 2, fascicle 1. Sarasota, Florida.

Dodson, Calaway H., and Piedad Marmol de Dodson. 1989. *Orchids of Ecuador. Icones Plantarum Tropicarum*, series 2, fascicle 5. St. Louis, Missouri: Missouri Botanical Garden.

Dodson, Calaway H., and Carlyle Luer. 2005. *Flora of Ecuador*. Vol. 76, Orchidaceae, *Genera AA–Cyrtidiorchis*. Saint Louis, Missouri: Missouri Botanical Garden.

Dodson, Calaway H., and Gustavo Romero. 1995. Revalidation of the genus *Benzingia* (Zygopetalinae: Orchidaceae). *Lindleyana* 10 (2): 74.

Dodson, Calaway H., and Roberto Vásquez. 1989. *Orchids of Bolivia. Icones Plantarum Tropicarum*, series 2, fascicles 3 and 4. St. Louis, Missouri: Missouri Botanical Garden.

Dressler, Robert L. 1981. *The Orchids: Natural History and Classification*. Cambridge, Massachusetts: Harvard University Press.

Dressler, Robert L. 1993a. *Field Guide to the Orchids of Costa Rica and Panama*. Ithaca, New York: Comstock Publishing Associates.

Dressler, Robert L. 1993b. *Phylogeny and Classification of the Orchid Family*. Portland, Oregon: Dioscorides Press, an imprint of Timber Press.

Dressler, Robert L. 2001. *Chondroscaphe. Orquideología* 22 (1): 16–22.

Dunsterville, G. C. K., and E. Dunsterville. 1974. *Huntleya* species of Venezuela. *Orchid Digest* 38 (3): 115–119.

Dunsterville, G. C. K., and Leslie A. Garay. 1959–1976. *Venezuelan Orchids Illustrated*. 6 vols. London: Andre Deutsch.

Dunsterville, G. C. K., and Leslie A. Garay. 1979. *Orchids of Venezuela: An Illustrated Field Guide*. 3 vols. Cambridge, Massachusetts: Botanical Museum of Harvard University.

Escobar R., Rodrigo. 1994. *Native Colombian Orchids*. Vols. 1–6. Medellín, Colombia: Compañía Litografica Nacional.

Fernández, César. 2003. *Orquídeas nativas del Táchira* (Táchira and its native orchids). Venezuela: Loteria del Táchira.

Fowlie, Jack A. 1966. *Huntleya fasciata. Orchid Digest* 30: 281.

Fowlie, Jack A. 1967. Some observations on the genus *Huntleya* and related genera. *Orchid Digest* 31: 279–280.

Fowlie, Jack A. 1968. An annotated checklist of the genus *Pescatorea. Orchid Digest* 32: 86–91.

Fowlie, Jack A. 1969a. An annotated checklist of the genus *Bollea. Orchid Digest* 33 (3): 100–103.

Fowlie, Jack A. 1969b. An annotated checklist of the genus *Warczewiczella. Orchid Digest* 33 (6): 224–231.

Fowlie, Jack A. 1971. *Chondrorhyncha helleri. Orchid Digest* 35: 170.

Fowlie, Jack A. 1984. A further contribution to an understanding of the genus *Huntleya. Orchid Digest* 48 (6): 221–225.

Garay, Leslie. 1969. El complejo *Chondrorhyncha. Orquideología* 4: 139–152.

Garay, Leslie. 1973. El complejo *Zygopetalum. Orquideología* 8 (1): 15–51.

Gerlach, Günther, and Tilman Neudecker. 1994. *Kefersteinia escobariana*: a new species from Ecuador. *Orquideología* 19 (3): 44–51.

Gloudon, Ancile, and C. Tobisch. 1995. *Orchids of Jamaica*. Kingston, Jamaica: University of the West Indies.

Hamer, Fritz. 1981. *Las Orquídeas de El Salvador*. San Salvador, El Salvador: Ministerio de Education.

Hamer, Fritz. 1982. *Orchids of Nicaragua. Icones Plantarum Tropicarum*, fascicle 13, part 1–6. Sarasota, Florida: Marie Selby Botanical Gardens.

Hammel, Barry E., Michael H. Grayum, Cecilia Herrera, and Nelson Zamora, eds. (in prep.) *Manual de Plantas de Costa Rica*. Monographs in Systematic Botany, vol. 6. St. Louis, Missouri: Missouri Botanical Garden.

Jenny, R. 1983. *Chondrorhyncha bicolor* Rolfe. *Die Orchidee* 34 (6): 315–316.

Lindley, John. 1837. *Maxillaria steelii. Edwards's Botanical Register* 23: 1986.

Neudecker, Tilman. 1982. *Pescatorea whitei. Orchidee (Hamburg)* 33 (4): 129.

Neudecker, Tilman, and Günther Gerlach. 2000. Rediscovery of the genus *Dodsonia*, and description of a new *Stenia* from Ecuador: *Stenia glatzii. Orquideología* 21 (3): 256–267.

Pridgeon, Alec M., editor. 1992. *The Illustrated Encyclopedia of Orchids*. Portland, Oregon: Timber Press.

Pridgeon, Alec M., Phillip J. Cribb, Mark W. Chase, Finn N. Rasmussen, eds. 1999. *Genera Orchidacearum* Vol. 1, *General Introduction*. Oxford, England: Oxford University Press.

Pupulin, Franco. 2001. Contributions toward a reassessment of Costa Rican

Zygopetalinae (Orchidaceae). 2. A reconsideration of *Chondrorhyncha estrellensis* Ames. *Brenesia* 55–56: 135–140.

Pupulin, Franco. 2005. *Vanishing Beauty: Native Costa Rican Orchids.* Vol. 1, *Acianthera–Kegeliella.* San José, Costa Rica: Universidad de Costa Rica.

Reichenbach, Heinrich Gustav. 1858–1900. *Xenia Orchidacea.* 4 vols. Leipzig: F. A. Brockhaus.

Rungius, C. 1998. Checkliste zu den Gattungen der Huntleyinae. *Die Orchidee* 49: 172–179, 211–219, 296–298.

Saunders, William W., ed. 1869. *Refugium Botanicum; or Figures and Descriptions from Living Specimens of Little Known or New Plants of Botanical Interest,* vol. 1. London: J. van Voorst.

Schultes, Richard E., and Leslie A. Garay. 1957–1959. On the validity of the generic name *Cochleanthes* Raf. *Botanical Museum Leaflets,* Harvard University, 18: 321–322.

Schweinfurth, Charles. 1959. *Orchids of Peru.* Chicago: Chicago Natural History Museum.

Senghas, Karlheinz, and G. Gerlach. 1992–1993. 59 Subtribus Huntleyinae. In F. R. R. Schlechter, *Die Orchideen* 1/B: 1620–1674. Berlin and Hamburg. Verlag Paul Parey.

Senghas, Karlheinz, and G. Gerlach. 1993a. 60 Subtribus Zygopetalinae. In F. R. R. Schlechter, *Die Orchideen* 1/B: 1674–1727. Berlin and Hamburg. Verlag Paul Parey.

Senghas, Karlheinz, and G. Gerlach. 1993b. 691 *Chondroscaphe.* In F. R. R. Schlechter, *Die Orchideen* 1/B (27): 1655. Berlin and Hamburg. Verlag Paul Parey.

Stearn, William T. 1995. *Botanical Latin.* Portland, Oregon: Timber Press.

Szlachetko, Darius L. 2003. *Senghasia,* eine neue Gattung der Zygopetaleae. *Journal für den Orchideenfreund* 10 (4): 332–344.

Szlachetko, Darius, and A. Romowicz. 2006. Notes sur le genre *Senghasia* Szlachetko (Orchidaceae, Huntleyinae). *Richardiana* 6: 180–182.

Urreta, Guillermo M. 2005. *Orquídeas de la Serranía del Baudó Chocó.* Bogotá, Colombia: Capitalina de Orquideología.

van der Pijl, L., and Calaway H. Dodson. 1966. *Orchid Flowers: Their Pollination and Evolution.* Coral Gables, Florida: University of Miami Press.

Vásquez, Roberto, and Calaway H. Dodson. 1982. *Orchids of Bolivia. Icones Plantarum Tropicarum,* series 1, fascicle 6. St. Louis, Missouri: Missouri Botanical Garden.

w^{3}TROPICOS. Missouri Botanical Garden, St. Louis. Published on the Internet at http://mobot.mobot.org/W3T/Search/vast.html.

Waldvogel, E. 1982. *Bollea hirtzii. Orchidee (Hamburg)* 33 (4): 143.

Whitten, W. Mark, Norris H. Williams, Robert L. Dressler, Günther Gerlach, and Franco Pupulin. 2005. Generic relationships of Zygopetalinae

(Orchidaceae: Cymbidieae): combined molecular evidence. *Lankesteriana* 5 (2): 87–107.

Williams, Louis O., Paul H. Allen, and Robert L. Dressler. 1980. *Orchids of Panama*. Monographs in Systematic Botany, vol. 4. St. Louis, Missouri: Missouri Botanical Garden.

World Checklist of Monocotyledons. The Board of Trustees of the Royal Botanic Gardens, Kew. Published on the Internet at http://www.kew.org/wcsp/monocots.

Index of Plant Names